A Primer for Modern Mathematics

A Primer for Modern Mathematics

Bernard W. Banks
California Polytechnic State University

WCB Wm. C. Brown Publishers
Dubuque, Iowa•Melbourne, Australia•Oxford, England

Book Team

Editor *Paula-Christy Heighton*
Developmental Editor *Jane Parrigin*
Production Editor *Audrey A. Reiter*
Designer *Barbara J. Hodgson*

Wm. C. Brown Publishers
A Division of Wm. C. Brown Communications, Inc.

Vice President and General Manager *Beverly Kolz*
Vice President, Publisher *Earl McPeek*
Vice President, Director of Sales and Marketing *Virginia S. Moffat*
National Sales Manager *Douglas J. DiNardo*
Marketing Manager *Julie Joyce Keck*
Advertising Manager *Janelle Keeffer*
Director of Production *Colleen A. Yonda*
Publishing Services Manager *Karen J. Slaght*
Permissions/Records Manager *Connie Allendorf*

Wm. C. Brown Communications, Inc.

President and Chief Executive Officer *G. Franklin Lewis*
Corporate Senior Vice President, President of WCB Manufacturing *Roger Meyer*
Corporate Senior Vice President and Chief Financial Officer *Robert Chesterman*

Cover design by Kay Fulton

Copyedited by Joseph Pomerance

A Times Mirror Company

Library of Congress Catalog Card Number: 93–72696

ISBN 0–697–21494–X

Printed in the United States of America by Wm. C. Brown Communications, Inc., 2460 Kerper Boulevard, Dubuque, IA 52001

10 9 8 7 6 5 4 3 2 1

Contents

A Foreword to the Reader

This book is about set theory, proof, the real number system, and the axiomatic method. In this foreword I would like to take the oportunity to explain why these topics were chosen. The principal aim of this book is to provide a basis for the further study of modern mathematics, and so the main topics of this book were chosen because they are at the foundation of mathematics.

Just prior to the beginning of the nineteenth century the content of mathematics was relatively easy to describe. There was plane and solid Euclidean geometry, algebra and the solution of equations, and the calculus. By the middle of the twentieth century the number of major areas of mathematical interest had multiplied into a veritable zoo of topics. To name a few, we have groups, rings, fields, vector spaces, Banach and Hilbert spaces, topological spaces, manifolds, varieties, homology and cohomology, differential and integral equations, etc. Yet in spite of this multiplicity, modern mathematics has a remarkable unity. The unity is provided by the fact that almost all the concepts of mathematics find their expression in the language of sets. As a simple example, the reader may recall how important the notions of a relation and function are in mathematics, and the reader may recall that these notions can be precisely defined in terms of sets of ordered pairs. Since set theory is fundamental in mathematics today, almost half of this book is devoted to this topic.

A second unifying theme in modern mathematics is that each area of mathematics is developed as a deductive science. This theme is not new in mathematics. Indeed, Euclid (ca. 300 B.C.) and the ancient Greek geometers established the deductive or axiomatic method over two thousand years ago. In brief, the method consists in the logical deduction of conclusions called theorems from a given list of assumptions called axioms or postulates. The reader may recall the method from a high school geometry course. Let me dwell for a moment on the need for this method in mathematics.

The Theorem of Pythagoras says that the area of the square constructed on the hypotenuse of a right angled triangle is equal to the sum of the areas of the squares constructed on the other two sides. Perhaps, like the ancient Egyptians and Babylonians, you have discovered this theorem by experimenting with a few right triangles. The following questions may have occurred to you. Will the theorem hold with any right triangle you pick, and if so, why? These questions urge us to search for a proof of the theorem. A proof of the theorem must satisfy us that there can be no right triangle for which the theorem fails. One way to do this would be to start from some geometrical facts that we believe to be true and construct a chain of propositions that ends with the Theorem of Pythagoras and has the property that any proposition in the chain is true if its predecessors are true. This is the deductive method in a nutshell.

It is worth noting that the deductive method would appear to be the only method available to establish the truth of propositions about infinite sets of objects such as the set of all right triangles. After all, we cannot check every triangle to see if the Theorem of Pythagoras holds.

As a second example consider the assertion that there are infinitely many prime numbers. The truth of this cannot be decided by any other means than a proof based on the properties of addition and multiplication of whole numbers. Again, we cannot check all whole numbers to see if they have proper factors or not. Indeed, any time we make assertions about infinite sets of objects deductive proof is usually the only way to establish the truth of the assertion.

The third major topic in this book is a rigorous axiomatic treatment of the real number system and its subsystems the natural numbers, integers, and rational numbers. The main reason for the inclusion of this topic is that the real number system forms the principal measuring stick of science and commerce, and it forms the foundation of the calculus and the rest of analysis. In addition, the development of the properties of the real numbers provides a good example of the axiomatic method.

The fourth major topic is the axiomatic method. The reason for including this topic is that each area of modern mathematics is, when consciously defined, axiomatic. That is to say, each area of mathematics is defined by a set of propositions or axioms given as a starting point from which further theorems are deduced. Thus, to a first approximation, to study the axiomatic method is to study the structure of modern mathematics – at least in its public or written aspect. The axiomatic method provides a serviceable definition of mathematics, but it must be remembered that mathematizing is a human activity, and, as such, cannot be fully described in any simple formula. Tennis is defined by the rules of the game, but the game of tennis is played by people and becomes much more than its set of rules.

The book begins in the first chapter with the development of the logic necessary for the expression of statements and for the employment of deduction and proof. In the second chapter the theory of sets is introduced, and the assumptions about sets are formulated precisely in the form of axioms. Set theory is further developed in Chapters 2 and 3 by introducing the reader to the basic methods of proof and applying them to the deduction of results from the axioms of set theory. So by the end of Chapter 3 the reader will have been introduced to the mathematical fundamentals of set theory and proof. At the same time the reader will have experienced the development of an area of mathematics in a thoroughly modern way. Chapter 4 develops the properties of the basic number systems axiomatically, and thus it provides the reader with a second model example of the axiomatic method. Chapter 5 is devoted to further study of the axiomatic method and the way in which it is employed to define mathematical structures. In particular, the question of the existence and uniqueness of mathematical structures defined by axioms is considered. Finally, in Chapter 6 we discuss and prove Zorn's lemma. The reason for doing this is that Zorn's Lemma plays a central role in the proof of many theorems in advanced mathematics.

A Note to Instructors

Chapters 1, 2, and 3 should be covered in order. However, section 3.6 in Chapter 3 (The Axiom of Choice) may be omitted until Chapter 6. In Chapter 4, the sections on base ten and decimal representations can be omitted and the student's prior knowledge relied on if time is short. Chapter 6 is independent of Chapters 4 and 5. The construction of the real numbers from the natural numbers in Chapter 5 is intended to add breadth to the student's knowledge, but it can be omitted without harm.

While the text material probably lends itself best to a lecture format, the exercises could be tackled in a small group setting. This is particularly true of the long and developmental exercise sets in Chapter 5.

Finally, Chapters 1–4 provide material for a one quarter course, and the complete book provides material for a semester course.

Acknowledgments

I would like to take this opportunity to thank Dr. Barbara Shabell and Dr. Harriet Lord for their many helpful criticisms. In particular, I am grateful to Dr. Shabell for reading the rough draft and making many useful suggestions. I am most grateful to Dr. Lord for testing the book in the classroom and for all of the cogent suggestions that she made in the course of the development of this book.

Chapter 1

The Elements of Mathematical Language

1.1 Propositions and Operations

The purpose of this book is to assist the reader in gaining admission to modern mathematics. The principal feature of modern mathematics is its insistence that its truths be deduced from clearly stated assumptions. Thus, modern mathematics has returned to the same standards that were established by Euclid (ca. 300 B.C.) over two thousand years ago. Therefore, first of all we need to develop the elements of mathematical language necessary to the conduct of deductive reasoning. Since deductive reasoning uses statements, we need to begin with that notion. The kind of statement we have in mind is a sentence that has the property of being true or false. For example, the sentence "John's car is red" asserts something that is true or false. In logic such declarative sentences are called propositions, and we shall follow this custom. Here are some more examples of propositions.

Examples

a) $2 < 3$. b) $2 < 0$
c) Mr. Lincoln was president. d) Sugar is sweet. •

Note that we shall indicate the end of definitions, examples, and proofs with the bullet (•).

New propositions can be generated from old by performing certain logical operations. The operations of importance to us are negation, conjunction, disjunction, implication, and equivalence.

Negation

If P is a proposition then the negation of P is the proposition, it is not the case that P. We denote this proposition by $-P$. Note that if P is true then $-P$ is false, and if P is false then $-P$ is true. This can be summarized neatly in a table.

P	$-P$
T	F
F	T

For $-P$ we often write: "not P".

Conjunction

If P and Q are propositions we can form a new proposition: $P\&Q$. This new proposition asserts that P and Q are both true, so $P\&Q$ is true only if both P and Q are true. Again a table expresses this neatly.

P	Q	$P\&Q$
T	T	T
T	F	F
F	T	F
F	F	F

Examples

a) $(2 < 3)$ & (π is irrational). (True)
b) $(2 < 3)$ & $(1 < 0)$. (False)
c) $(2 = 3)$ & $(5 = -1)$. (False) •

For $P\&Q$ one often writes: "P and Q", or "$P \wedge Q$". We call $P\&Q$ the conjunction of P and Q.

Disjunction

Given propositions P and Q, the new proposition $P \vee Q$ is false only if both P and Q are false. It is true otherwise. This new proposition is called the disjunction of P and Q, and it has the same meaning as the legal barbarism "and/or". Here is the table.

P	Q	$P \vee Q$
T	T	T
T	F	T
F	T	T
F	F	F

Examples

a) $(2 < 3) \vee (4 < 5)$. (True)
b) $(2 < 3) \vee (5 < 4)$. (True)
c) $(3 < 2) \vee (5 < 4)$. (False) •

We often write for $P \vee Q$: "P or Q".

We can extend the definition of conjunction and disjunction to any number of propositions. Thus, $P_1 \& P_2 \& \ldots \& P_n$ is true only if all of $P_1 \ldots P_n$ are true. Similarly, $P_1 \vee P_2 \vee \ldots \vee P_n$ is false only if all $P_1 \ldots P_n$ are false.

Implication

By an implication we mean a proposition of the form: "if P then Q". A common symbolism for implication is $P \Rightarrow Q$. Based on our ordinary usage of $P \Rightarrow Q$ we would agree that it is true if P and Q are true and false if P is true and Q false, but it is not immediately clear how to assign the truth values T or F in the other two cases. Consider when $P \Rightarrow Q$ is false. In ordinary usage this is the case if P is true and Q is false. So it is reasonable to take $P \& -Q$ as the negation of $P \Rightarrow Q$ in ordinary usage. If we accept this then we must agree that if P is false and Q is true then $P \Rightarrow Q$ is true, and if P is false and Q is false then $P \Rightarrow Q$ is true. If the reader is still uneasy then the table below should be accepted as the definition of $P \Rightarrow Q$.

P	Q	$P \Rightarrow Q$
T	T	T
T	F	F
F	T	T
F	F	T

Examples

a) If $2 < 3$ then $3 < 4$. (True)
b) If $2 < 3$ then $4 < 3$. (False)
c) If $3 < 2$ then $3 < 4$. (True)
d) If $3 < 2$ then $4 < 3$. (True) •

In the expression "$P \Rightarrow Q$" we call P the antecedent and Q the conclusion. Thus, an implication is false only if the antecedent is true and the conclusion is false.

The reader may still feel uneasy about examples c) and d) above. Part of the discomfort may arise from thinking of $P \Rightarrow Q$ as saying P causes Q. From this point of view the choice of truth value in the last two lines of the table does indeed seem arbitrary, but this is not its meaning in logic. Perhaps the following example will ease the discomfort.

Consider the implication: If it is a warm day then I will go to the beach. Suppose you agree that the implication is false provided it expresses a lie. Now suppose it is a cold day and I go to the beach. Is the implication a lie? No! It says nothing about what I may or may not do on cold days. So in this case the implication must be accepted as true. The last line of the table must be assigned a T on the basis of the same reasoning.

The reader should note the following terminology. We call $Q \Rightarrow P$ the converse of $P \Rightarrow Q$.

Equivalence

By the proposition P iff Q we shall mean $(P \Rightarrow Q)\&(Q \Rightarrow P)$. For P iff Q we read "P if and only if Q", and refer to it as an equivalence. Clearly P iff Q is true provided both P and Q are true or both are false. The reader is asked to construct the table for equivalence in the exercises.

Terminology

The tables we have constructed are called truth tables. Given a proposition P, assigning true or false (T or F) to P is called assigning a truth value. A proposition such as $(P \Rightarrow Q)\&(R \vee S)$ is called a compound proposition, and the individual letter propositions P, Q, R, and S are called atomic propositions.

Here is the truth table for $P \Rightarrow (Q \vee R)$ where we have three atomic propositions.

P	Q	R	$Q \vee R$	$P \Rightarrow (Q \vee R)$
T	T	T	T	T
T	T	F	T	T
T	F	T	T	T
T	F	F	F	F
F	T	T	T	T
F	T	F	T	T
F	F	T	T	T
F	F	F	F	T

Tautologies and Contradictions

It is an interesting fact that certain compound propositions are true regardless of the truth value of the atomic propositions.

Example

Consider the important proposition $[P \& (P \Rightarrow Q)] \Rightarrow Q$ called modus ponens by logicians. Here is its truth table.

P	Q	$P \Rightarrow Q$	$P \& (P \Rightarrow Q)$	$[P \& (P \Rightarrow Q)] \Rightarrow Q$
T	T	T	T	T
T	F	F	F	T
F	T	T	F	T
F	F	T	F	T

Notice that the last column contains all Ts. This is an example of a tautology since it is a proposition that is true regardless of the truth values of the atomic propositions. •

The importance of modus ponens in deductive logic is that if you know that P is true, and you know that $P \Rightarrow Q$ is true, then you can conclude that Q is true also. This is reflected in the first row of the above table.

A contradiction is a compound proposition that is false regardless of the truth values of its atomic propositions. Clearly the negation of any tautology is a contradiction, but here is the truth table of a very useful one.

S	$-S$	$S \& -S$
T	F	F
F	T	F

We shall introduce other tautologies and contradictions as we need them.

Exercises for Section 1.1

Determine the truth or falsity of the following propositions.

1. $-(2 < 3)$.
2. –(The Earth is flat).
3. If $2 > 3$ then $7 > 8$.
4. The Earth is flat or water is wet.
5. If $2 < 3$ then $-(7 > 8$ or $\sqrt{2}$ is rational).
6. $\sqrt{2}$ is irrational, and (π is irrational or 1/2 is irrational).
7. If $[2 < 3$ and (if $2 < 3$ then $3 < 4)]$ then $3 < 4$.
8. $2 < 3$ iff $3 < 4$.
9. $2 < 3$ iff $4 < 3$.

Construct truth tables for the following, and determine which are tautologies or contradictions or neither.

10. P iff Q.
11. $P \vee -P$.
12. $(P \& Q) \Rightarrow P$.
13. $[(P \vee Q) \& -Q] \Rightarrow P$.
14. $[(P \Rightarrow Q) \& (Q \Rightarrow R)] \Rightarrow (P \Rightarrow R)$.
15. $-\{[(P \Rightarrow Q) \& -Q] \Rightarrow -P\}$.
16. $(P \Rightarrow Q)$ iff $(-Q \Rightarrow -P)$.
17. $-(P \vee Q)$ iff $(-P \& -Q)$.

1.2 Open Sentences and Quantifiers

Open Sentences

Many sentences that occur in mathematics have the syntactical form of a proposition, but they are not propositions. For example, "x is red", or "$x < 2$". We do not regard such sentences as propositions because, unless x is specified, we cannot say that the sentences are true or false. If, however, x is replaced with the name of an object then the sentences become propositions. For example, $x < 2$ cannot be said to be true or false, but if x is replaced by say 4 we get a false proposition. Sentences that contain a variable or placeholder, which when replaced by the name of an object become propositions, are called open sentences or predicates. We shall use expressions such as $P(x)$, $Q(t)$, or $R(y)$ to denote open sentences.

Examples

a) $x^2 \geq 0$. b) t is an insect. c) $y + 3 = 7$. •

It may already have occurred to the reader to ask if the variable in an open sentence can be replaced by the name of anything. At first sight it may appear so, but example a) above puts the lie to this. In a) above try replacing x by Donald Duck. The sentence (Donald Duck)$^2 \geq 0$ makes no sense, and so it is not a proposition. In order to avoid such nonsense we must restrict the candidates that are to replace x to a collection of objects for which the sentence makes sense. This collection or set of candidates to replace the variable is called the universe for that variable. Since this is our first contact with the concept of a set it is appropriate to comment on it at this time. We shall return to the concept of a set in much greater detail in Chapter 2.

Sets

Language is full of words that express the set concept. For example, we have: collection, group, flock, covey, school, aggregate, crowd, etc. In mathematics the notion of a set is as fundamental as the elements of logic, and yet no definition can be given that is not circular. For example, to say a set is a collection is to shift the problem to defining what a collection is. Does this mean that mathematics is a house built on sand? The answer to such a basic question is, as might be expected, no and yes! It is no because, however murky our perception of the concept of a set may be, we can agree on some propositions that express properties that sets have. We can then develop our theories on the basis of these properties. So while we do not see the concept of a set with perfect clarity, we can state some of its properties clearly. The answer is yes because we do not know if the propositions we clearly set out are consistent. They might lead to a contradiction. However, the properties of sets that we shall lay out have not yielded a contradiction so far.

The fundamental idea associated with the notion of a set is membership. Symbolically this is expressed as follows:

$$x \in A.$$

In this expression A is a set, and we read the expression as: "x is in A", or "x belongs to A", or "x is an element of A", etc. In Chapter 2, we carefully state the properties of sets in terms of the relation of membership. For now, our intuitve notion of a set as a collection will suffice.

If a set has a finite membership it can be described by simply listing its members. The standard notation for this is:

$$\{a_1, a_2, \ldots, a_n\}.$$

Quantifiers

There are two fundamental ways in which an open sentence such as $x^2 \geq 0$ can be converted into a proposition. Let $P(x)$ be an open sentence. To convert this into a proposition we first specify a universe for the variable x. We denote it by U_x. We shall use the convention of subscripting U with a variable to denote the universe of that variable. Given $P(x)$ and the universe U_x, a pair of alternatives presents itself. Either $P(x)$ is true for every replacement of x by members of U_x or it is not. Thus, "$P(x)$ is true for every choice of x in U_x" is a proposition, and we denote it by $\forall(x)P(x)$. Therefore, $\forall(x)P(x)$ is false if there is an object a in U_x such that $P(a)$ is false. We call $\forall(x)$ a universal quantifier quantifying the variable x.

Another pair of alternatives is either $P(x)$ is true for some element in U_x, or it is not. Thus, "there exists an x in U_x such that $P(x)$ is true" is a proposition. It is often denoted by $\exists(x)P(x)$. We call $\exists(x)$ an existential quantifier, quantifying the variable x.

Examples

1) Let U_x be the set of real numbers. Then,
   a) $\forall(x)(x^2 \geq 0)$ is true, and b) $\forall(x)(x + 2 = 0)$ is false.
2) a) Let U_x be the set $\{1, 2, 3\}$, then $\forall(x)(x > 0)$ is true.
   b) Let U_x be the set $\{-1, 0, 1\}$, then $\forall(x)(x > 0)$ is false.
3) a) Let U_x be the set $\{1, 2, 3\}$, then $\exists(x)(x > 2)$ is true.
   b) Let U_x be the set $\{-1, 0, 1\}$, then $\exists(x)(x > 2)$ is false.

Notice in these examples how the universe affects the truth or falsity of the statement.

If an element c in U_x is such that $P(c)$ is false, then $\forall(x)P(x)$ is false. In this case c is called a counter-example. Let U_x be the set of real numbers. Consider the proposition $\forall(x)(x$ has a real square root). Then we would say that -1 is a counter-example to this universally quantified statement. Counter-examples play a very important role in the development of a field in mathematics. In fact this little counter-example can be viewed as provoking the development of the complex numbers.

The ideas just discussed can easily be extended to open sentences or predicates in more than one variable. We shall discuss the case of two variables, and the reader will see how to extend these ideas to more than two variables.

One can easily think of sentences that are open in two variables. For example, "x is the brother of y", or "$x > y$". Clearly such expressions are neither true nor false until the variables x and y are replaced by the names of objects from appropriate universes U_x and U_y. As before, if $P(x,y)$ is a sentence open in x and y, we must restrict x and y to universes for which $P(x,y)$ will be meaningfull. Also as before, we can employ quantifiers to convert $P(x,y)$ into a proposition. There are several possibilities, and we treat enough of them for the principles to become clear.

It will be convenient to quantify $P(x,y)$ in stages. Consider $\forall(y)P(x,y)$. This is open in the variable x because if x is replaced by the name of an object in U_x, say a, then $\forall(y)P(a,y)$ is a proposition. Thus, $\forall(x)\forall(y)P(x,y)$ is a proposition. Indeed, it is true precisely when $P(x,y)$ is true for all replacements of x and y by names of elements in U_x and U_y, respectively. Also it is clear that $\forall(x)\forall(y)P(x,y)$ iff $\forall(y)\forall(x)P(x,y)$.

Similarly, $\exists(y)P(x,y)$ is open in x, and so $\exists(x)\exists(y)P(x,y)$ is a proposition. Indeed, it is clear that quantifying a variable reduces the number of variables in which the expression is open by one. The meaning of $\exists(x)\exists(y)P(x,y)$ is straightforward. The proposition $\exists(x)\exists(y)P(x,y)$ is true only if there are elements, say a in U_x and b in U_y, such that $P(a,b)$ is true.

Examples

Let U_x and U_y be the set of real numbers.

a) $\forall(x)\forall(y)(x^2+y^2 \geq 0)$ is true.
b) $\forall(x)\forall(y)(x + y = 0)$ is false.
c) $\exists(x)\exists(y)(x^2 + y^2 = 0)$ is true.
d) $\exists(x)\exists(y)(x^2 + y^2 < 0)$ is false.•

When we mix the universal and existential quantifiers the meaning becomes a little more subtle, and the reader should take special care in this. What do we mean by $\forall(x)\exists(y)P(x,y)$, and $\exists(y)\forall(x)P(x,y)$? The proposition $\forall(x)\exists(y)P(x,y)$ is true if to each x in U_x there corresponds a b_x in U_y such that $P(x,b_x)$ is true. Contrast this with the condition under which $\exists(y)\forall(x)P(x,y)$ is true. This is true if there is some object b in U_y such that $\forall(x)P(x,b)$ is true. And this is true if $P(x,b)$ is true for all replacements of x by members of U_x. In this case the b "works" for all x, whereas $\forall(x)\exists(y)P(x,y)$ is true means to each x there corresponds a b_x such that $P(x,b_x)$ is true.

Consider the following concrete example. Let U_x be the set of all people, and let U_y be the set of all hats. Then the sentence "For every x there is a y such that x owns y" is very different from the sentence "There is a y such that for all x, x owns y". A little reflection shows that

$$\exists(y)\forall(x)P(x,y) \Rightarrow \forall(x)\exists(y)P(x,y),$$

but not vice versa.

Examples

Let U_x and U_y be the set of real numbers.

a) $\forall(x)\exists(y)(x + y = 0)$ is true.
b) $\forall(x)\exists(y)(xy = 1)$ is false.
c) $\exists(y)\forall(x)(x + y = 0)$ is false.
d) $\exists(y)\forall(x)(xy = 0)$ is true. •

We close this section with a mathematical structure that provides useful examples now and in the future. Consider the set $\mathbf{Z}_n = \{0, 1, 2, \ldots, n-1\}$ where n is a positive integer. If x and y are in $\mathbf{Z}_n$ then we define addition and multiplication as follows. We define $x \oplus_n y$ to be the remainder when $x + y$ is divided by n. Similarly, $x \otimes_n y$ is the remainder when xy is divided by n. So, for example, $3 \oplus_6 5 = 3$, $2 \otimes_6 3 = 0$ and $2 \otimes_6 2 = 4$. Here $\mathbf{Z}_6 = \{0, 1, 2, 3, 4, 5\}$.

Example

With $\mathbf{Z}_4$, $x \oplus_4 y$ and $x \otimes_4 y$ as described above, let $U_x = U_y = \mathbf{Z}_4$. Now consider the proposition $\forall(x)\exists(y)(x \oplus_4 y = 0)$. It is true as we now demonstrate. For $x = 0$, $y = 0$ is the corresponding element such that $x \oplus_4 y = 0$. For $x = 1$, $y = 3$ is the element such that $x \oplus_4 y = 0$. For $x = 2$, $y = 2$ is what is needed. For x equal to 3, 1 is the corresponding value of y. •

Exercises for Section 1.2

Decide if the following propositions are true or false and give the reason.

1. Let U_x be the set of real numbers.
   a) $\forall(x)(x^2 + 1 > 0)$.
   b) $\forall(x)(x^3 > 1)$.
   c) $\forall(x)(\text{if } x > 0 \text{ then } x^3 > 0)$.
   d) $\forall(x)(\text{if } x^2 < 0 \text{ then } x = 1)$.
   e) $\exists(x)(x > 0 \;\&\; x < 0)$.

2. Let $U_x = \mathbf{Z}_5$.
   a) $\forall(x)(x \oplus_5 0 = x)$.
   b) $\forall(x)(x \oplus_5 2 \neq 0)$.
   c) $\forall(x)(x \otimes_5 1 = x)$.
   d) $\exists(x)(x \otimes_5 1 = 1)$.

3. Let $U_x = U_y = \mathbf{Z}_7$.
   a) $\forall(x)\exists(y)(x \oplus_7 y = 0)$.
   b) $\exists(y)\forall(x)(x \oplus_7 y = 0)$.

4. Let $U_x = U_y = \{1, 2, 3, 4, 5, 6\}$.
   a) $\forall(x)\exists(y)(x \otimes_7 y = 1)$.
   b) $\exists(y)\forall(x)(x \otimes_7 y = 0)$.

5. Let $U_x = U_y = \{1, 2, 3, 4, 5, 6, 7\}$.
   a) $\forall(x)\exists(y)(x \otimes_8 y = 1)$.
   b) $\exists(y)\forall(x)(x \otimes_8 y = 0)$.

6. a) With $U_x = U_y = \mathbf{Z}_n$ for n = 2, 3, 4, 5, and 9 check the truth or falsity of $\forall(x)(\text{if } x \neq 0 \text{ then } \exists(y)(x \otimes_n y = 1))$.
   b) For what values of n do you think the proposition is true?

1.3 Logical Validity and Negations

Let R stand for a compound logical expression of the type we have been considering. If R is composed only of atomic propositions, we say that R is logically valid provided that R is a tautology. In the case where R is composed of quantified predicates, we say R is logically valid provided it is true no matter what meaning the predicate letters might be given.

Examples

The reader can easily construct truth tables to verify that the following are logically valid.

a) $-(P \vee Q)$ iff $(-P \& -Q)$. b) $-(P \& Q)$ iff $(-P \vee -Q)$.

c) $[P \& (P \Rightarrow Q)] \Rightarrow Q$. d) $[(P \vee Q) \& -Q] \Rightarrow P$. •

Examples

a) $\forall(x)(P(x) \& Q(x))$ iff $[\forall(x)P(x) \& \forall(x)Q(x)]$ is logically valid. Indeed, regardless of what P or Q might be, if $\forall(x)(P(x) \& Q(x))$ is true then $P(x)$ and $Q(x)$ are true no matter what x is, and conversely.

b) $[\forall(x)P(x) \vee \forall(x)Q(x)] \Rightarrow \forall(x)(P(x) \vee Q(x))$ is logically valid, for if the antecedent is true then $\forall(x)P(x)$ or $\forall(x)Q(x)$ is true, in which case $P(x) \vee Q(x)$ is true for each x. Note that the converse

$$\forall(x)(P(x) \vee Q(x)) \Rightarrow [\forall(x)P(x) \vee \forall(x)Q(x)]$$

is not logically valid. •

If (R iff S) is logically valid we say that S is a logical equivalent of R, or that (R iff S) is a logical equivalence. We shall find it useful to have certain standard logical equivalents of negations of quantified statements, so we develop them below.

Consider $-\forall(x)P(x)$. If it is true then $\forall(x)P(x)$ is false, so there is an object b in the universe U_x such that $P(b)$ is false. Therefore, $\exists(x)(-P(x))$ is true. It is also easy to see that if $\exists(x)(-P(x))$ is true then so is $-\forall(x)P(x)$. Thus,

$$-\forall(x)P(x) \text{ iff } \exists(x)(-P(x)),$$

is a logical equivalence.

If $-\exists(x)P(x)$ is true then $\exists(x)P(x)$ is false, so there is no x for which $P(x)$ is true. So for each x, $-P(x)$ is true. Thus, $\forall(x)(-P(x))$ is true. By reversing this argument we see that

$$\forall(x)(-P(x)) \text{ iff } -\exists(x)P(x),$$

is logically valid.

Similar consideration of the meaning of the quantifiers involved shows that

$$-\forall(x)\forall(y)P(x,y) \text{ iff } \exists(x)\exists(y)(-P(x,y)),$$

and

$$-\exists(x)\exists(y)P(x,y) \text{ iff } \forall(x)\forall(y)(-P(x,y)),$$

is logically valid.

Now suppose $-\forall(x)\exists(y)P(x,y)$ is true. Then $\forall(x)\exists(y)P(x,y)$ is false, and so there is no relation $x \to b_x$ from U_x to U_y such that for each x, $P(x,b_x)$ is true. So for some x in U_x, call it c, there is no corresponding b_c such that $P(c,b_c)$ is true. So for all y in U_y, $P(c,y)$ is false. Thus, $\exists(x)\forall(y)(-P(x,y))$ is true. Reversing this argument shows that

$$-\forall(x)\exists(y)P(x,y) \text{ iff } \exists(x)\forall(y)(-P(x,y)),$$

is a logical equivalence.

By reasoning in a similar manner the reader can see that

$$-\exists(x)\forall(y)P(x,y) \text{ iff } \forall(x)\exists(y)(-P(x,y)),$$

is a logical equivalence.

The following rule suggests itself. To negate a quantified statement move the negation sign inside and exchange all the quantifiers. This rule is valid, but it would take more time than we care to spend to set up the machinery to prove it in complete generality.

A Word About Equality

What is meant when we write $a = b$? We mean that the symbols "a" and "b" stand for the same thing. In other words, $a = b$ means a and b are identical. Generally, if $a = b$ we may exchange a for b or vice versa in a logical expression without affecting the truth of the expression. It is true that this general rule can get one in trouble, but we shall not run into any such cases. The interested reader may find additional information on this point in the reference at the end of the chapter.

Exercises for Section 1.3

1. Show that the following are logically valid.
   a) $-(P \Rightarrow Q)$ iff $(P\&-Q)$. b) $[(P \Rightarrow Q) \;\&\; -Q] \Rightarrow -P$.

2. Find logical equivalents of the following negations.
   a) $-(P\vee(Q\vee R))$. b) $-(P\&(Q\&R))$. c) $-[(P\&Q) \Rightarrow R]$.

3. Find logical equivalents of the following negations. Let U_x be the set of real numbers. Also, decide if the statement is true.

a) $-\forall(x)(x^2 > 0)$.
b) $-\forall(x)(x = 2 \vee x = 3)$.
c) $-\forall(x)(\text{if } x = 1 \text{ then } x^2 = 1)$.
d) $-\exists(x)(\text{if } x < 0 \text{ then } x^2 = 0)$.
e) $-\exists(x)(x = 2 \;\&\; x < 0)$.

4. Find logical equivalents of the following negations. Let U_x, and U_y be the set $\mathbf{Z}_6$. Also, decide if the statement is true.
a) $-\forall(x)\exists(y)(x \otimes_6 y = 1)$.
b) $-\forall(x)\exists(y)(\text{if } x \neq 0 \text{ then } x \otimes_6 y = 1)$.
c) $-\exists(x)\forall(y)(x \oplus_6 y = 0)$.
d) $-\exists(x)\forall(y)(x \otimes_6 y = 0)$.

5. Find a logical equvalent of

$$-\forall(\varepsilon)\exists(\delta)\forall(x)\{\text{if } 0 < |x - a| < \delta \text{ then } |f(x) - L| < \varepsilon\}.$$

Reference

P. Suppes, *Introduction to Logic* (D. Van Nostrand, Princeton, New Jersey, 1964)

Chapter 2

Set Theory and Proof

2.1 What Is a Proof?

We start with a definition of the word "proof" that is broad enough to cover the many uses that it has in mathematics. By a proof of a proposition we mean a demonstration that the proposition is true. The following examples show the need for such a broad definition.

The presentation of a truth table is a proof of the truth of a tautology. The sentence $3 \times 4 = 12$ is a proof that 12 is a composite number. The function $f(x) = |x|$ shows that there are functions that are continuous at zero but not differentiable there. In geometry, drawing diagrams can be central to the demonstration of a truth. Finally, we mention the form of proof that is of greatest importance in this book, namely deduction.

Each area of mathematics is defined by stating some propositions or axioms that are assumed to be true of the mathematical structure being defined. The object is to discover what else is true given that the axioms are true. If $\mathbb{A}$ represents the conjunction of the axioms then the object is to find propositions Q such that $\mathbb{A} \Rightarrow Q$ is true. We call Q a theorem of the mathematical area. In order for Q to be accepted as a theorem the truth of the implication $\mathbb{A} \Rightarrow Q$ must be proved. We now give a sketch of the way in which deductive proof achieves this end.

In a deductive proof the central structure is a sequence of propositions or predicates with the property that if the predecessors of a proposition or predicate in the sequence were true then that proposition or predicate would be true. Thus, if $P_1, P_2, \ldots, P_n$ is such a sequence then, if P_1 is true then so is P_n. Therefore, $P_1, P_2, \ldots, P_n$ constitutes a proof that $P_1 \Rightarrow P_n$. Thus, if $\mathbb{A} = P_1$ and $Q = P_n$ then we also have a proof that $\mathbb{A} \Rightarrow Q$ is true.

If our theorem to be proved involves open sentences (predicates) and quantifiers the situation is more complicated. For example, suppose the theorem to be proved is $\forall(x)Q(x)$. The method is this. Drop the universal quantifier, and then think of x as the name of any object in the universe of x. We can then think of $Q(x)$ as a proposition. This process is called universal specification. We now prove $Q(x)$ as before. That is, we find a sequence $\mathbb{A}, P_1, \ldots, P_n, Q(x)$ such that if $\mathbb{A}$ is true then P_1 is true, and if P_1 is true P_2 is true, and so on. Thus, we have a proof that $\mathbb{A} \Rightarrow Q(x)$ is true. Finally, if the proof "$\mathbb{A}, P_1, \ldots, P_n, Q(x)$" does not depend on the choice of x we conclude that $\mathbb{A} \Rightarrow \forall(x)Q(x)$ is true. This last step is called the universal generalization.

Quantified statements of many different forms will appear in our proofs repeatedly. We shall not try to give the formal rules for the treatment of each of the many possibilities now. Instead, we shall introduce the necessary methods as we need them. We shall see that these methods are naturally motivated by the meaning of the quantifiers and the main goal of demonstrating that an expression of the form $\mathbb{A} \Rightarrow Q$ is true.

We close this section with the following observation. We often want to prove theorems of the form $P \Rightarrow Q$. So we must show that $\mathbb{A} \Rightarrow (P \Rightarrow Q)$ is true. But

$$[(\mathbb{A} \ \& \ P) \Rightarrow Q] \Rightarrow [\mathbb{A} \Rightarrow (P \Rightarrow Q)]$$

is a tautology. The reader should construct the truth table. Therefore, to prove $\mathbb{A} \Rightarrow (P \Rightarrow Q)$ we take P as a temporary axiom called the temporary premise and prove Q.

2.2 Two Proof Strategies

Contraposition Proofs

To prove a theorem of the form $P \Rightarrow Q$ it is enough to prove a logical equivalent. The reader should construct the truth table for the tautology

$$(P \Rightarrow Q) \text{ iff } (-Q \Rightarrow -P).$$

We call $-Q \Rightarrow -P$ the contrapositive of $P \Rightarrow Q$. Since $-Q \Rightarrow -P$ is logically equivalent to $P \Rightarrow Q$ it is enough to prove $-Q \Rightarrow -P$. To prove this, $-Q$ is taken as the temporary premise and $-P$ is deduced. This method of proof is called the method of contraposition.

The reader should be very careful to note that $P \Rightarrow Q$ is **not** logically equivalent to its converse $Q \Rightarrow P$. Thus, in proving $P \Rightarrow Q$ **never** take Q as a temporary premise.

Contradiction Proofs

To prove a theorem $\boldsymbol{P}$ of $\mathbb{A}$ by contradiction one takes $-\boldsymbol{P}$ as the temporary premise and then one tries to deduce a contradiction of the form $S\&-S$. If we are successful, we have shown that $[\mathbb{A}\&-P] \Rightarrow (S\&-S)$ is true. But $S\&-S$ is false, so $\mathbb{A}\&-P$ must be false. Therefore, if $\mathbb{A}$ is true then $-P$ must be false, and so P must be true. This shows that $\mathbb{A} \Rightarrow P$ is true and P is a theorem of $\mathbb{A}$.

The rest of this chapter is devoted to developing the area of set theory using the methods of proof we have described. In doing so, the generalities described above will become concrete.

2.3 Set Theory

If a set is a collection of objects it seems easy to define all manner of sets. It seems that for any predicate $P(x)$ we get a set by considering the collection of all objects x for which $P(x)$ is true. The reader is probably already familiar with the notation used for such sets. It is the so called set builder notation,

$$\{x \mid P(x)\}.$$

We read this as " the set of all x such that $P(x)$".

However, the unrestricted use of this means of defining sets immediately leads to a contradiction. Suppose for any set A we define $P(A)$ to mean $A \notin A$. Now form $B = \{A \mid P(A)\} = \{A \mid A \notin A\}$. Now if B is a set then $B \in B$, or $B \notin B$. Suppose $B \in B$ then by definition of B, $B \notin B$. And if $B \notin B$, then by definition of B, $B \in B$. This is the famous Russel paradox. The only way out is to assume that either B does not exist, or that whatever B is it is not a set. It is usual to say that B exists but is not a set. We regard B as a collection of a higher order, and the name class is reserved for such large things. Another example of a class is $\{A \mid A \text{ is a set}\}$. Thus, the class of all sets must not be treated as a set lest we fall into a contradiction again.

In order to avoid contradictions, limitations must be placed on the use of the set building construction. The axioms set out below provide such limitations.

Basic Axioms for Sets

The following form a provisional collection of axioms for sets. We shall add to the collection in due course.

Before stating the axioms a word about the universe for the variables is in order. If A is a set, the expression $x \in A$ is meaningful no matter what x may be. So in set theory, where $x \in A$ is the only predicate, we may take the universe to be the class of all mathematical objects. Sets are mathematical objects, so this class contains the subclass of all sets. Thus, the class of all mathematical objects is unimaginably large, and is certainly not a set. We denote this class by U.

The axioms consist in asserting that the following constructions define sets.

1) $\phi = \{x \mid x \neq x\}$ is a set.
2) $\forall(x)(\{t \mid t = x\}$ is a set). For a given x we denote this set by $\{x\}$ and call it singleton x.
3) $\forall(A)\forall(B)$(if A and B are sets then $\{x \mid x \in A$ or $x \in B\}$ is a set). The set $\{x \mid x \in A$ or $x \in B\}$ is denoted by $A \cup B$, and it is called the union of A and B.
4) $\forall(A)\forall(B)$(if A and B are sets then $\{x \mid x \in A$ and $x \in B\}$ is a set). The set $\{x \mid x \in A$ and $x \in B\}$ is denoted by $A \cap B$, and it is called the intersection of A and B.
5) $\forall(A)\forall(B)$(if A and B are sets then $\{x \mid x \in A$ and $x \notin B\}$ is a set). The set $\{x \mid x \in A$ and $x \notin B\}$ is denoted by $A - B$, and it is called the relative complement of B in A.

Definition 2.3.1

Let A and B be sets. We say $A \subset B$ iff $\forall(x)$(if $x \in A$ then $x \in B$). In this case we say A is a subset of B. •

We now continue with the axioms.

6) $\forall(A)\forall(B)$(if A and B are sets then $A = B$ iff $A \subset B$ and $B \subset A$).
7) Let $P(x)$ be a predicate, and let A be any set on which $P(x)$ makes sense.Then $\{x \mid x \in A$ and $P(x)\}$ is a set.
8) Let A be any set. Then $\{B \mid B \subset A\}$ is a set. This set is the set of all subsets of A , and it is called the power set of A. We shall denote the power set by $\mathcal{P}(A)$.

Two remarks about the axioms and definition are in order. First, the reader should note that a definition is not an axiom but an agreement to use words or symbols in place of some existing expression. It is an agreement to use one expression as a shorthand for another. Thus, $A \cup B$, $A \cap B$, etc are the shorthand expressions for the sets declared to exist in the axioms.

Second, the reader may feel that axiom 6) is unnecessary. To be sure, $A = B$ means A and B denote the same set by the meaning of equality. So, it follows directly that $\forall(x)$(if $x \in A$ then $x \in B$) and $\forall(x)$(if $x \in B$ then $x \in A$), but the converse does not follow from the meaning of equality. A golf club and a yacht club could have the same membership, but they are not the same club.

Axiom 6) says that, unlike clubs, sets are completely determined by their memberships.

We now provide some examples to illustrate the axioms. In the next section our first theorem will show how the axioms can be used to justify the existence of sets defined by the listing method that we use below.

Examples

Let $A = \{1, 2, 3, 4, 5\}$ and $B = \{2, 4, 6, 7\}$. Then $A \cup B = \{1, 2, 3, 4, 5, 6, 7\}$, $A \cap B = \{2, 4\}$, and $A - B = \{1, 3, 5\}$. Now A is not a subset of B, but if $C = \{1, 2\}$ then $C \subset A$. Finally, we give an example of axiom 7). We have, $\{x \mid x \in A \text{ and } x \text{ is even}\} = \{2, 4\}$. •

Examples of power sets will be given after a few more results about sets have been developed.

Exercises for Section 2.3

1. Let $A = \{3, 2, 5, 7\}$ and $B = \{3, 7, 4, 1\}$. Write down $A \cup B$, $A \cap B$, and $A - B$. Is A a subset of B? Explain!

2. Which of the following statements are true? Explain!
   a) $1 \in \Phi$.
   b) $1 \in \{1\}$.
   c) $\Phi \in \{\Phi\}$.
   d) $\{\Phi\} \in \Phi$.
   e) $\{\Phi, \{\Phi\}\} \in \{\{\Phi, \{\Phi\}\}\}$.
   f) $\{1\} \subset \{1, 4, 8\}$.
   g) $\Phi \not\subset \{1, 4, 8\}$.
   h) $\{1\} \in \{1, \{1\}, 2, \{1, 4, 5\}\}$.
   i) $\{\Phi, \{\Phi\}\} \subset \{\Phi, 2, 3, \{\Phi\}\}$

2.4 Some Elementary Proofs

Throughout the rest of this chapter and the rest of this book we shall assume that the axioms of set theory are true. Therefore we will speak of the theorems of set theory as being true.

We advise readers to adopt the following questioning "mental set" throughout the rest of this book. Ask of all assertions in theorems and proofs "is it true?". Let no proposition pass that you cannot justify. Constantly ask the question "why?", and do not move on until you are satisfied.

We begin our discussion of proofs with the proof of a theorem that does not involve an implication. We shall organize our initial proofs in two columns. On the left is the proof, and on the right are listed justifications and remarks.

Theorem 2.4.1

$\forall(x)\forall(y)(\{x\} \cup \{y\}$ is a set)

We remark that the first step in this proof is to drop the universal quantifiers $\forall(x)\forall(y)$ and choose x and y as arbitrary elements of the universe. In this way we may think of the predicate

"$\{x\} \cup \{y\}$ is a set"

as a proposition. Recall that this procedure is called universal specification. The idea is that if the proposition without quantifiers can be shown to be true for an arbitrary choice of variables, then it must be true for any values of the variables. If so, then the quantifiers can be replaced and the quantified statement is true. The replacement is called universal generalization. Observe the process in the following proof.

Proof:	Justifications:
1. Let x and $y \in U$.	1. Universal specification.
2. $\{x\}$ is a set.	2. True by axiom 2.
3. $\{y\}$ is a set.	3. True by axiom 2.
4. $\{x\} \cup \{y\}$ is a set.	4. True by axiom 3.
5. $\forall(x)\forall(y)(\{x\} \cup \{y\}$ is a set).	5. Universal generalization.

The reader should take note of lines 1 and 5. In line 1 we choose x and y to stand for arbitrarily chosen objects in our universe of mathematical objects. Lines 2–4 show that for these objects $\{x\} \cup \{y\}$ is a set. Nothing in lines 2–4 depends on anything except that x and y are in U. So in line 5) we can say that for every x and y in U, $\{x\} \cup \{y\}$ is a set. Put another way, lines 2–4 eliminate the possibility that there exist a and b in U such that $\{a\} \cup \{b\}$ is not a set.

The general rule is this; if one arrives at a sentence $P(x_1, \ldots, x_n)$ in a proof where the derivation of this sentence depends only on the x_i being in their universes, U_i, then one can universally generalize on these variables.

These points are covered again in the theorems that follow. In order that this beginning discussion not be too cluttered with quantifiers, let A, B, C etc. denote sets in the following theorems.

Theorem 2.4.2

$A \subset A \cup B$.

By Definition 2.3.1 $A \subset A \cup B$ means $\forall(x)$(if $x \in A$ then $x \in A \cup B$) . Therefore, we shall prove the latter. Because of the implication the proof will employ a temporary premise.

Proof:	Justification:
1. Let $x \in U$.	1. Universal specification.
2. Suppose $x \in A$.	2. Assumed true as the temporary premise.
3. $x \in A$ or $x \in B$.	3. If $x \in A$ is true then so is $x \in A$ or $x \in B$.
4. $x \in A \cup B$.	4. True by the definition of $A \cup B$ in axiom 3.
5. if $x \in A$ then $x \in A \cup B$.	5. From lines 2 and 4.
6. $\forall(x)$(if $x \in A$ then $x \in A \cup B$).	6. Universal generalization.
7. $A \subset A \cup B$.	7. By Def. 2.3.1

In line 1 we simply choose an arbitrary object in the universe and call it "x". Line 2 is the temporary premise in this proof. If we assume the temporary premise is true, then lines 3 and 4 are true. So lines 2 through 4 form a proof that "if $x \in A$ then $x \in A \cup B$" is true. This is stated in line 5. Observe that the truth of line 5 follows for any choice of x, so in line 6 the universal generalization is made. Finally, line 7 is true since lines 6 and 7 are equivalent by Definition 2.3.1.

Consider the following theorems.

Theorem 2.4.3

$A \cap B \subset A$.

We shall prove that $\forall(x)$(if $x \in A \cap B$ then $x \in A$) is true. So after the universal specification we shall take $x \in A \cap B$ as the temporary premise.

Proof:	Justification:
1. Let $x \in U$.	1. Universal specification.
2. Suppose $x \in A \cap B$.	2. Temporary premise. Assume it is true.
3. $x \in A$ and $x \in B$.	3. True by axiom 4 and the definition of $A \cap B$.
4. $x \in A$.	4. True by the meaning of "and".
5. If $x \in A \cap B$ then $x \in A$.	5. True, since lines 2 – 4 form a proof of this.
6. $\forall(x)$(if $x \in A \cap B$ then $x \in A$).	6. True by universal generalization.

7. $A \cap B \subset A$.	7. True by Def. 2.3.1.

Theorem 2.4.4

$$A - (B \cup C) = (A - B) \cap (A - C)$$

Since this theorem is an equality, axiom 6) requires that we prove that

$A - (B \cup C) \subset (A - B) \cap (A - C)$ and

$(A - B) \cap (A - B) \subset A - (B \cup C)$.

The proof is in two parts. This theorem is one of De Morgan's laws.

Part 1.

Proof:	Justification:
1. Let $x \in U$.	1. Universal specification.
2. Suppose $x \in A - (B \cup C)$.	2. Temporary premise assumed true.
3. $x \in A$ and $x \notin B \cup C$.	3. True by the meaning of $A - (B \cup C)$.
4. $x \in A$.	4. True by the meaning of "and".
5. $x \notin B \cup C$.	5. True by the meaning of "and".
6. $-(x \in B \cup C)$.	6. Equivalent to line 5.
7. $-(x \in B$ or $x \in C)$.	7. True by the meaning of union.
8. $x \notin B$ and $x \notin C$.	8. True since $-(P$ or $Q)$ iff $(-P$ and $-Q)$ is a tautology.
9. $x \in A$ and $x \notin B$.	9. True from lines 4 and 8.
10. $x \in A$ and $x \notin C$.	10. True from lines 4 and 8.
11. $x \in (A - B)$ and $x \in (A - C)$.	11. From lines 9 and 10.
12. $x \in (A - B) \cap (A - C)$.	12. From the meaning of intersection.
13. If $x \in A - (B \cup C)$ then $x \in (A - B) \cap (A - C)$.	13. True from lines 2 – 12.
14. $\forall(x)$(if $x \in A - (B \cup C)$ then $x \in (A - B) \cap (A - C)$).	14. Universal generalization.
15. $A - (B \cup C) \subset (A - B) \cap (A - C)$.	15. True by Def. 2.3.1.

In the second part we omit the justifications, and we invite the reader to jot down the justifications for each line. The reader should also notice that the proof of the second part is shorter, since a few steps have been combined. This points up the fact that the amount of detail included in a proof is often a matter of taste.

Part 2.

Proof:
1. Suppose $x \in (A - B) \cap (A - C)$.
2. $x \in (A - B)$, and $x \in (A - C)$.
3. $x \in A$.
4. $x \notin B$ and $x \notin C$.
5. $x \notin (B \cup C)$. Why?
6. $x \in A - (B \cup C)$.
7. $(A - B) \cap (A - C) \subset A - (B \cup C)$.

In the second part the universal specification and generalization have been omitted. Lines 1 through 6 give a proof that (if $x \in (A - B) \cap (A - C)$ then $x \in A - (B \cup C)$) is true. Line 7 then follows easily.

The amount of detail one includes in a proof usually depends on the audience to whom it is addressed. But one should always be prepared to return to the complete detail of Theorems 2.4.1–2.4.3 if necessary. In the next theorem we shall drop the column of justifications and omit some details, but we will add remarks parenthetically as we see the need.

Theorem 2.4.5

If $A \subset B$ then $A \subset A \cap B$.

Since this theorem is an implication, we shall take $A \subset B$ as a temporary premise and then prove $A \subset A \cap B$. To do this we shall need to take $x \in A$ as an additional temporary premise. We shall then prove that $x \in (A \cap B)$.

Proof:
1. Suppose $A \subset B$. (This is a temporary premise)
2. Let $x \in A$.
3. $x \in B$. (Since $A \subset B$.)
4. $x \in A$ and $x \in B$.
5. $x \in (A \cap B)$.
6. If $x \in A$ then $x \in (A \cap B)$. (True from lines 2 through 5)
7. $A \subset A \cap B$. (Universal generalization and definition of $\subset$)
8. If $A \subset B$ then $A \subset A \cap B$.

Theorem 2.4.5 has the form $P \Rightarrow \forall(x)(P(x) \Rightarrow Q(x))$. Lines 1 through 5 show that $(\mathbb{A} \& P \& P(x)) \Rightarrow Q(x)$ is true. P and $P(x)$ are temporary premises. Since $(\mathbb{A} \& P \& P(x)) \Rightarrow Q(x)$ is true so is

$$(\mathbb{A} \& P) \Rightarrow (P(x) \Rightarrow Q(x)).$$

Since the argument does not depend on the choice of x

$$(\mathbb{A} \& P) \Rightarrow \forall(x)(P(x) \Rightarrow Q(x))$$

is true by universal generalization. Finally, we conclude that

$$\mathbb{A} \Rightarrow [P \Rightarrow \forall(x)(P(x) \Rightarrow Q(x))]$$

is true.

Definition 2.4.6

Let X be a set and A a subset of X. We write A^c for $X - A$. We call A^c the complement of A. •

In the next theorem let A and B be subsets of a fixed set X.

Theorem 2.4.7

If $A \subset B$ then $B^c \subset A^c$.

The proof of this theorem, like that of the preceding theorem, will begin with $A \subset B$ as the primary temporary premise. Assuming that $A \subset B$ is true, we shall prove that $B^c \subset A^c$ is true.

Proof:
1. Suppose $A \subset B$.
2. Let $x \in B^c$.
3. $x \notin B$.
4. $x \notin A$. (Otherwise $x \in B$.)
5. $x \in A^c$.
6. $B^c \subset A^c$.

Branching

Every now and then we come to a line in a proof of the form:

n. $P \vee Q$.

This causes us to pause, since, while we know that "$P \vee Q$" is true, we cannot assert either one is true by itself. Thus, we seem unable to proceed. We have to apply the method of branching.

Suppose our object is to show that S is true. We know that P is true or Q is true, so in turn we prove that $P \Rightarrow S$ and $Q \Rightarrow S$ are true. Thus if P or Q or both are true S must be true since we have covered all possible cases.

A more formal justification of the branching procedure is that

$$[(P \Rightarrow S) \;\&\; (Q \Rightarrow S)] \Rightarrow [(P \vee Q) \Rightarrow S]$$

is a tautology. We apply the branching procedure in the next theorem.

Theorem 2.4.8

If $A \subset B$ then $A \cup B \subset B$.

Proof:
1. Suppose $A \subset B$.
2. Let $x \in A \cup B$.
3. $x \in A$ or $x \in B$. (Now we must branch. Our goal is $x \in B$.)
4. a) Suppose $x \in A$.
   b) $x \in B$. {By line 1.}
5. a) Suppose $x \in B$.
   b) $x \in B$. (Trivial, but we put the step in to complete the procedure.)
6. $x \in B$.
7. If $x \in A \cup B$ then $x \in B$.

Thus we can conclude that the theorem is true.

Exercises for Section 2.4

In the following exercises let A, B, and C be subsets of a set X.

1. Prove the following in complete detail. Provide a justification for each line.
   a) $\forall(x)\forall(y)\forall(z)((\{x\} \cup \{y\}) \cup \{z\}$ is a set).
   b) $A \cap B \subset A \cup B$.
   c) $(A \cap B)^c = A^c \cup B^c$.
   d) $(A \cup B)^c \subset A^c \cap B^c$.
   e) If $A \cap B = A$ then $A \subset B$.

2. Prove the following in the more informal style of theorems 2.4.7 and 2.4.8.
   a) $A \cap (B \cup C) = (A \cap B) \cup (A \cap C)$. (Branching?)
   b) $A \cup (B \cap C) = (A \cup B) \cap (A \cup C)$. (Branching?)
   c) $(A^c \cup B)^c = A \cap B^c$.
   d) $A^{cc} = A$.

3. Prove the following in the informal style of 2.4.7 and 2.4.8.
   a) If $C \subset A \cap B$ then $C \subset A$.
   b) If $A \subset B$ and $B \subset C$ then $A \subset C$.
   c) If $A \cup B = B$ then $A \subset B$.
   d) If $A \cap B = A$ then $A \subset B$.
   e) If $A \subset B$ and $C^c \subset B^c$ then $C^c \subset A^c$.

4. If you have not already proved 3 b), do so. Give a detailed explanation of your proof, modeling it on the explanation following the proof of Theorem 2.4.5.

2.5 Contraposition and Contradiction Proofs

In this section contraposition and contradiction proofs are demonstrated. In the process, we show how to deal with existential quantifiers.

Contraposition

Recall that $(P \Rightarrow Q)$ iff $(-Q \Rightarrow -P)$ is a tautology. The method of proof by contraposition consists in proving $(-Q \Rightarrow -P)$ in place of $(P \Rightarrow Q)$. We shall reprove Theorem 2.4.7, so let A and B be subsets of a set X.

Theorem 2.5.1

If $A \subset B$ then $B^c \subset A^c$.

Proof:

1. Suppose $B^c \not\subset A^c$. (This is the temporary premise for the contraposition proof.)
2. $-\forall(x)($ if $x \in B^c$ then $x \in A^c$). (Definition 2.3.1.)
3. $\exists(x)(x \in B^c \ \& \ x \notin A^c)$. (Equivalent to line 2.)
4. Let p be an element such that $p \in B^c$ and $p \notin A^c$. (This is called existential specification.)
5. $p \notin B$ and $p \in A$.
6. $\exists(x)(x \in A \ \& \ x \notin B)$. {This is called existential generalization.}
7. $-\forall(x)($ if $x \in A$ then $x \in B$).
8. $A \not\subset B$.

Since the proof has shown that "if $B^c \not\subset A^c$ then $A \not\subset B$" is true, we can conclude that the theorem is true by the method of contraposition.

As far as the actual proof is concerned, there is only one new point to make. Line 3 says there exists an object such that $x \notin A^c \ \& \ x \in B^c$ is true when x is replaced by the name of that object. In line 4 we give that object a name by calling it "p". Thus existential specification is just the process of supplying such names. In line 5 we see that there is an object, namely, p, such that $x \in A \ \& \ x \notin B$ is true when x is replaced by p. Thus, line 6 is true by the meaning of $\exists(x)$.

Contradiction

Recall that the strategy of proof by contradiction is based on the implication $(\mathbb{A}\&-P) \Rightarrow (S\&-S)$. To prove that P is true we prove that this implication is true. But $S\&-S$ is false, and we have assumed $\mathbb{A}$ is true, so $-P$ must be false, and so P must be true.

Theorem 2.5.2

$\forall(x)(x \notin \phi)$.

Proof:
1. Suppose $-\forall(x)(x \notin \phi)$. (This is the temporary contradiction premise.)
2. $\exists(x)(x \in \phi)$.
3. Let p be such that $p \in \phi$. (Existential specification.)
4. $p \neq p$. (Axiom 1.)
5. $p = p$. (Always true by the meaning of equality.)
6. $p \neq p \;\&\; p = p$. (Contradiction $S\&-S$.)

We conclude that $\forall(x)(x \notin \phi)$ is true by the method of contradiction. In view of this theorem we call ϕ the empty set since it has no members.

We now illustrate the method of proof by contradiction in two more theorems.

Theorem 2.5.3

$\phi \subset A$.

Proof:
1. Suppose $\phi \not\subset A$. (Contradiction premise.)
2. $-\forall(x)$(if $x \in \phi$ then $x \in A$).
3. $\exists(x)(x \in \phi \;\&\; x \notin A)$.
4. Let p be such that $p \in \phi \;\&\; p \notin A$.
5. $p \in \phi$.
6. $p \neq p$.
7. $p \neq p \;\&\; p = p$. (You could get a contradiction using Theorem 2.5.2.)

So we conclude that $\phi \subset A$. In fact the proof is independent of the choice of the set A, so we can universally generalize to get

$$\forall(A)(\text{if } A \text{ is a set then } \phi \subset A).$$

Thus, the empty set is a subset of every set.

Theorem 2.5.4

If $A \cap B = A$ then $A \subset B$.

Proof:

1. Suppose $A \cap B = A$ and $A \not\subset B$. (Equivalent to the negation of the theorem.)
2. $A \not\subset B$.
3. $\exists(z)(\, z \in A$ and $z \notin B\,)$. (Skipped a step, and the variable can be anything.)
4. Let p be such that $p \in A$ and $p \notin B$.
5. $p \in A \cap B$. (From $A \cap B = A$ and $p \in A$.)
6. $p \in A$ & $p \in B$.
7. $p \in B$ & $p \notin B$.

We conclude that the theorem is true.

Exercises for Section 2.5

1. Use the method of contraposition proof to prove the following. Model your proofs on that of Theorem 2.5.1.
   a) If $C \subset A \cap B$ then $C \subset A$.
   b) If $A \subset B$ then $A \cup B = B$.
   c) If $A \subset B$ and $B \subset C$ then $A \subset C$.
   d) If $A \cup B = B$ then $A \subset B$.

2. Let A and B be subsets of a set X. Prove, using contraposition, that if $A = B^c$ then $A^c = B$.

3. Prove the following by the method of contradiction. Model your proofs on the examples in the text. Let A and B be subsets of a set X.
   a) $A \subset A$.
   b) $B \subset A \cup A^c$.
   c) $A \cap A^c = \phi$.
   d) If $A \cup B = B$ then $A \subset B$. (Contradiction not contraposition.)
   e) If $A \subset B$ and $B \subset C$ then $A \subset C$.

2.6 Informal Proofs

In this section we give examples of the informal style of proof that is most common in mathematics. We shall adopt this style throughout the rest of the book. We begin by reproving some theorems we have already proved.

Theorem 2.6.1

If $A \subset B$ then $A \subset A \cap B$.

Proof:
Suppose $A \subset B$. Let $x \in A$. Then, $x \in B$. So $x \in A$ and $x \in B$. So $x \in A \cap B$. Thus, $A \subset A \cap B$. •

Note that we drop the numbering of propositions, and we connect them together with such words as "then", "so", etc. This provides for easy reading, and acts as a reminder that the proposition follows from some of its predecessors.

Theorem 2.6.2

For all A, if A is a set then $\phi \subset A$.

Proof:
Let A be a set. Suppose $\phi \not\subset A$. Then there is a p such that $p \in \phi$. So $p \neq p$. But that is a contradiction. Thus, $\phi \subset A$. •

Theorem 2.6.3

$$A - (B \cup C) = (A - B) \cap (A - C).$$

Proof:
Let $x \in A - (B \cup C)$. Then $x \in A$, and $x \notin B \cup C$. So, $x \notin B$ and $x \notin C$. Therefore, $x \in A - B$, and $x \in A - C$. So $x \in (A - B) \cap (A - C)$.

Let $x \in (A - B) \cap (A - C)$. Then, $x \in A - B$, and $x \in A - C$. So, $x \in A$, $x \notin B$, and $x \notin C$. Thus, $x \in A$, and $x \notin B \cup C$. So, $x \in A - (B \cup C)$. •

Note that, because of the equality, the proof is in two parts, and these parts are presented as paragraphs.

We conclude these examples with a theorem we have not proved before.

Theorem 2.6.4

If $A \cap B = \phi$ and $C \subset A$ then $B \cap C = \phi$.

Proof:
We prove this by contradiction. Suppose $A \cap B = \phi, C \subset A$, and $B \cap C \neq \phi$. Then there is a p such that $p \in B \cap C$. So $p \in B$, and $p \in C$. Thus, $p \in A$ (Why?). So $p \in A \cap B$, and so $p \in \phi$. But that is impossible. •

In deciding on the amount of detail you should include in an informal proof you should be guided by the following rule. If you make an assertion in a proof and you cannot answer the question "why?" in a few words, you probably do not understand your own proof, and so you need to add detail.

Exercises for Section 2.6

Provide informal proofs of a selection of the exercises in Sections 2.4 and 2.5.

2.7 Families of Sets

Beginners in set theory often have difficulty with the idea that sets are objects in our universe, and so we can have sets of sets. In this very brief section we try to nip this difficulty in the bud. At the same time we fulfill our promise to give examples of power sets.

The set axioms alow us to create all manner of sets. For example ϕ is a set, and so $\{\phi\}$, $\{\phi,\{\phi\}\}$, and $\{\{\phi\}, \{\phi,\{\phi\}\}\}$ are sets. In fact, they are sets of sets.

Given the set $A = \{1, 2, 3\}$, the axiom of the power set assures us of the existence of the power set $\mathcal{P}(A)$. This is the set of all subsets of A. Thus, $\mathcal{P}(A) = \{\phi,\{1\},\{2\},\{3\},\{1, 2\},\{1, 3\},\{2, 3\},\{1, 2, 3\}\}$. Note that we needed to know that the empty set is a subset of every set before we could construct $\mathcal{P}(A)$. It is reasonable to ask if there are any other subsets of A like ϕ that we have missed. Suppose $S \subset A$. We have two alternatives, $S = \phi$, or $S \neq \phi$. If $S = \phi$ we have counted it. If $S \neq \phi$, then it can only contain a combination of the elements 1, 2, or 3. We have listed all possibilities.

Sets of sets can be created in many ways using the set builder axiom. For example assume the real numbers form a set. Call it $\mathbb{R}$. Then the open intervals $(-x,x)$ such that $x > 0$ are sets. So

$$\{(-x,x) \mid x \in \mathbb{R} \text{ and } x > 0\}$$

is a set of sets.

When speaking of a set of sets one often uses the terminology "family of sets" or "collection of sets".

The following examples should help clarify these ideas.

Examples

The statements made in the following are true. Be sure you know why!

1) $\phi \subset \phi$, $\phi \notin \phi$, $\phi \in \{\phi\}$, $\phi \notin \{\{\phi\},\{\phi,\{\phi\}\}\}$.

Let $A = \{a, b\}$.

2) $\phi \in \mathcal{P}(A)$, $\{a\} \in \mathcal{P}(A)$, $A \in \mathcal{P}(A)$. •

We close this section with a simple theorem about power sets.

Theorem 2.7.1

For any sets A and B, if $A \subset B$ then $\mathcal{P}(A) \subset \mathcal{P}(B)$.

Proof:
Let A and B be sets with $A \subset B$. Suppose $S \in \mathcal{P}(A)$, then $S \subset A$. So, $S \subset B$ (Why?). So, $S \in \mathcal{P}(B)$. •

Exercises for Section 2.7

1. Let $A = \{1, 2, 3\}$. Decide if the following are true or false. Explain your answers.
   a) $\phi \subset A$. b) $\phi \in \mathcal{P}(A)$. c) $\{1\} \subset A$.
   d) $\{1\} \in \mathcal{P}(A)$. e) $\phi \subset \mathcal{P}(A)$.

2. Prove that for any sets A and B:
   a) $\mathcal{P}(A \cap B) = \mathcal{P}(A) \cap \mathcal{P}(B)$.
   b) $\mathcal{P}(A) \cup \mathcal{P}(B) \subset \mathcal{P}(A \cup B)$, but that equality is not possible in general.

3. Let $(a,b) = \{x \mid x \in \mathbb{R} \text{ and } a < x < b\}$, i.e., (a,b) is an open interval of reals. Let $\mathcal{F} = \{(1,2), (2,5), (0,1)\}$ and $\mathcal{G} = \{(0,2), (2,4)\}$. Find
   a) $\mathcal{F} \cap \mathcal{G}$ and b) $\mathcal{F} \cup \mathcal{G}$.

2.8 Union and Intersection of a Family of Sets

There are many occasions in mathematics when there is a need to take the union or intersection of more than two sets. The following axioms alow this.

Axiom for Arbitrary Unions (Axiom 9)

Let $\mathcal{F}$ be a family of sets. Then

$$\bigcup \mathcal{F} = \{x \mid \exists(A)(A \in \mathcal{F} \text{ and } x \in A)\},$$

is a set. We call this set the union of the family $\mathcal{F}$. This union is the set of all x such that x is in some set in $\mathcal{F}$. •

Axiom for Arbitrary Intersections (Axiom 10)

Let $\mathcal{F}$ be a family of sets wth $\mathcal{F} \neq \phi$ (See Exercise 6). Then

$$\bigcap \mathcal{F} = \{x \mid \forall(A)(\text{if } A \in \mathcal{F} \text{ then } x \in A)\},$$

is a set. We call this set the intersection of the family $\mathcal{F}$. This intersection is the set of all x such that x is in every set A in $\mathcal{F}$. •

Example

Let $\mathcal{F} = \{\{1\}, \{1, 2, 3, 6\}, \{1, 5, 2\}\}$. Then, $\bigcup \mathcal{F} = \{1, 2, 3, 5, 6\}$, and $\bigcap \mathcal{F} = \{1\}$. •

We conclude this section by illustrating proofs involving arbitrary unions and intersections. In these theorems the reader should note the way in which the families are defined using the set builder construction. In as much as

$$\mathcal{F} = \{A \mid A \in \mathcal{F}\},$$

it is consistent to write $\bigcap \{A \mid A \in \mathcal{F}\}$ for $\bigcap \mathcal{F}$.

Theorem 2.8.1 (A De Morgan Law)

Let X be a set, and let $\mathcal{F}$ be a family of subsets of X. Then

$$\bigcup \{X - A \mid A \in \mathcal{F}\} = X - \bigcap \{A \mid A \in \mathcal{F}\}.$$

Proof:

Suppose $x \in \bigcup \{X - A \mid A \in \mathcal{F}\}$. Then $x \in X - A_0$ for some $A_0 \in \mathcal{F}$. So $x \in X$, and $x \notin A_0$. So $x \notin \bigcap \{A \mid A \in \mathcal{F}\}$. (Why?) Hence $x \in X - \bigcap \{A \mid A \in \mathcal{F}\}$. (Why?)

Let $x \in X - \bigcap \{A \mid A \in \mathcal{F}\}$, then $x \in X$, and $x \notin \bigcap \{A \mid A \in \mathcal{F}\}$. But then $x \notin A_0$ for some $A_0 \in \mathcal{F}$. (Why?) Thus, $x \in X - A_0$.
So $x \in \bigcup \{X - A \mid A \in \mathcal{F}\}$. •

Theorem 2.8.2

Let C be a set and $\mathcal{F}$ a family of sets. Then

$$C \cap \bigcup\{A \mid A \in \mathcal{F}\} = \bigcup\{C \cap A \mid A \in \mathcal{F}\}.$$

Proof:

Let $x \in C \cap \bigcup\{A \mid A \in \mathcal{F}\}$. Then $x \in C$, and $x \in \bigcup\{A \mid A \in \mathcal{F}\}$. Thus, $x \in A_0$ for some A_0 in $\mathcal{F}$. Thus, for some A_0 in $\mathcal{F}$, $x \in C \cap A_0$. So $x \in \bigcup\{C \cap A \mid A \in \mathcal{F}\}$.

Conversely, if $x \in \bigcup\{C \cap A \mid A \in \mathcal{F}\}$, then for some A_0 in $\mathcal{F}$, $x \in C \cap A_0$. Then, $x \in C$ and $x \in A_0$ for some A_0 in $\mathcal{F}$. Therefore, $x \in C$, and $x \in \bigcup\{A \mid A \in \mathcal{F}\}$ (Why?). Thus, $x \in C \cap \bigcup\{A \mid A \in \mathcal{F}\}$. •

In the following theorem we assume any needed properties of the real numbers. In particular, we use the fact that given any real number $a > 0$, there is a positive integer n such that $1/n < a$.

Theorem 2.8.3

Let $\mathbb{R}$ be the set of real numbers. Let $\mathbf{N}$ be the set of positive integers. Let $n \in \mathbf{N}$ and $(1/(n+1),1) \subset \mathbb{R}$ be the open interval from $1/(n+1)$ to 1. Then

$$\bigcup\{(1/(n+1),1) \mid n \in \mathbf{N}\} = (0,1).$$

Proof:

Let $x \in \bigcup\{(1/(n+1),1) \mid n \in \mathbf{N}\}$. Then $x \in (1/(n+1),1)$ for some n in $\mathbf{N}$. So $0 < 1/(n+1) < x < 1$, and so $x \in (0,1)$.

Conversely, if $x \in (0,1)$, then $0 < x < 1$, and so there is an n in $\mathbf{N}$ such that $1/(n+1) < x < 1$. So, $x \in \bigcup\{(1/(n+1),1) \mid n \in \mathbf{N}\}$. (Why?) •

Exercises for Section 2.8

1. Prove the following. Let X be a set, and let $\mathcal{F}$ be a family of subsets of X. Then

$$\bigcap\{X - A \mid A \in \mathcal{F}\} = X - \bigcup\{A \mid A \in \mathcal{F}\}.$$

2. Prove the following. Let C be a set and $\mathcal{F}$ a family of sets. Then

$$C \cup \bigcap\{A \mid A \in \mathcal{F}\} = \bigcap\{C \cup A \mid A \in \mathcal{F}\}.$$

3. With the notation as in Theorem 2.8.3, prove that

$\bigcap\{(0,1+1/n) \mid n \in \mathbf{N}\} = (0,1]$. (Right closed interval.)

4. With the notation as in Theorem 2.8.3, prove that

$$\bigcap\{(1/(n+1),1) \mid n \in \mathbf{N}\} = (\tfrac{1}{2},1).$$

5. If $\mathcal{F} = \phi$ prove that $\bigcup\{A \mid A \in \mathcal{F}\} = \phi$.

6. In the axiom for arbitrary intersections we required that $\mathcal{F} \neq \phi$. Suppose that we allow $\mathcal{F} = \phi$ in this axiom. Then what happens?

References

P. Halmos, *Naive Set Theory* (Van Nostrand Reinhold Co., New York, 1960)
P. Suppes, *Introduction to Logic* (D. Van Nostrand, Princeton, New Jersey, 1964)

Chapter 3

Relations and Functions

In this chapter, we develop the fundamental concepts of relations and functions. These concepts are second only in importance to the concept of a set in mathematics. Indeed, relations and functions are defined in terms of sets.

3.1 Relations

When we think of two things as being related together we first of all think of the things, and then we express some kind of property they have. For example, "Bernard Banks is the brother of Gillian Ross" expresses a relationship between two people. But this is only an instance of the general relationship of brotherhood which can be expressed in the open sentence or predicate "x is the brother of y". Thus, relations are predicates in two variables. Here are two more examples: "$x \leq y$", and "$y = x^2$".

It is interesting to note that the pairs of objects that make "x is the brother of y" true, also make "x is male, and x and y have the same parents" true. The same may be said for the two sentences "$x \leq y$" and "$x + 2 \leq y + 2$". The sentences in these pairs satisfy the following definition of equivalence. We say two predicates $P(x,y)$ and $Q(x,y)$ are equivalent if

$$\forall(x)\forall(y)(P(x,y) \text{ iff } Q(x,y))$$

is true.

Do we regard equivalent predicates as defining the same or different relations? For example, are "$x \leq y$" and "$x + 2 \leq y + 2$" the same? This is a matter of taste, but in mathematics we prefer to think of equivalent predicates as defining the same relationship. We now describe how this is achieved.

Given that $P(x,y)$ and $Q(x,y)$ are equivalent, it is clear that the set of pairs of objects (a,b) with $a \in U_x$ and $b \in U_y$ for which $P(a,b)$ is true is the same set of pairs (a,b) for which $Q(a,b)$ is true. Thus, the equivalence of predicates can be expressed in terms of a set of pairs of objects. We shall regard this set of pairs for which the equivalent predicates are true as the relation that the equivalent predicates define. To make this precise we need some definitions.

Definition 3.1.1

Let a and b be members of the class U. We call (a,b) an ordered pair, and we require that for two ordered pairs (a,b) and (x,y), $(a,b) = (x,y)$ iff $a = x$ and $b = y$. •

Exercises 5 and 6 below show how the concept of an ordered pair can be defined within the framework of the set theory of Chapter 2. However, no harm will come if we regard an ordered pair as a list of two things in order. The latter notion can be extended to ordered lists of any finite length. An ordered list of length n is denoted by $(a_1, a_2, \ldots, a_n)$, and it is called an n-tuple.

Definition 3.1.2

Let A and B be sets. We define $A \times B = \{(x,y) \mid x \in A \text{ and } y \in B\}$. We call $A \times B$ the cartesian or cross product of A and B. •

That the cross product of two sets is a set can be accepted as an axiom, or it can be proved on the basis of the previous axioms and the set theoretic definition of an ordered pair given in the exercises below.

We now define relations as sets of ordered pairs.

Definition 3.1.3

Let A and B be sets. If R is a subset of $A \times B$ we say R is a relation from A to B. If $A = B$ we say R is a relation on A. •

Clearly, open sentences in two variables can be used to define relations. For example, given the sentence $P(x,y)$ we can define

$$R = \{(x,y) \mid P(x,y)\} \subset U_x \times U_y.$$

Given a relation $R \subset A \times B$, is there a predicate that can be used to define it? Yes! Set $P(x,y) = [(x,y) \in R]$. We shall often employ the notation xRy for $(x,y) \in R$. In that case we shall read xRy as "x is R-related to y" or "x is related to y by R".

The reason for taking the set point of view as primary is that it founds the idea of a relation in the theory of sets which we have already established. Thus relations are to be treated as members of the universal class U. As such, relations can be treated on the same footing as any other set. It makes relations mathematical objects rather than sentences.

We now define the domain and range of a relation.

Definition 3.1.4

Let $R \subset A \times B$ be a relation. We define

$$\mathrm{Dom}(R) = \{x \mid x \in A, \text{ and } \exists(y)(y \in B, \text{ and } (x,y) \in R)\}, \text{ and}$$
$$\mathrm{Ran}(R) = \{y \mid y \in B, \text{ and } \exists(x)(x \in A, \text{ and } (x,y) \in R)\}.$$

We call Dom(R) the domain of R, and we call Ran(R) the range of R. •

We now give examples of the concepts that have been introduced in this section.

Example

Let $A = \{1, 2, 3\}$ and $B = \{2, 5, 7, 8\}$. Then $R = \{(1,2), (1,5), (2,7)\}$ is a relation from A to B. We have $\mathrm{Dom}(R) = \{1, 2\}$, and $\mathrm{Ran}(R) = \{2, 5, 7\}$. •

If $\mathbb{R}$ is the set of real numbers then we can picture $\mathbb{R} \times \mathbb{R}$ as the coordinate plane. In this way relations from $\mathbb{R}$ to $\mathbb{R}$ can be visualised as is shown in the next examples.

Example

Let $R = \{(x,y) \mid x^2 + y^2 = 1\} \subset \mathbb{R} \times \mathbb{R}$. Then R is a relation from $\mathbb{R}$ to $\mathbb{R}$. This relation can be thought of as a subset of the plane as shown in the figure.

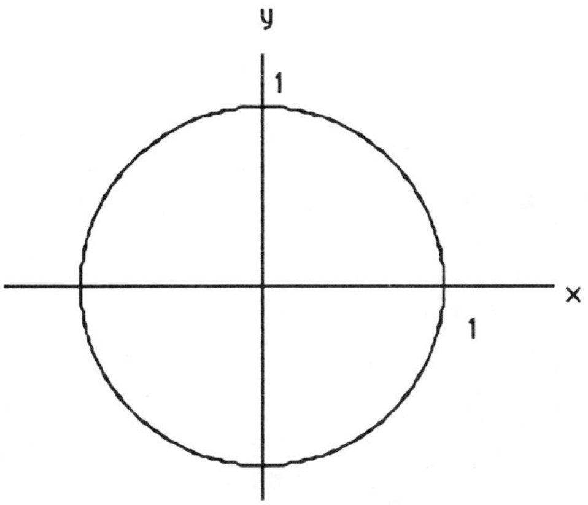

Clearly $\mathrm{Dom}(R) = [-1,1]$, and $\mathrm{Ran}(R) = [-1,1]$. •

Example

Let $R = \{(x,y) \mid x \leq y\} \subset \mathbb{R} \times \mathbb{R}$ then R is the region above and including the line $y = x$ in the figure.

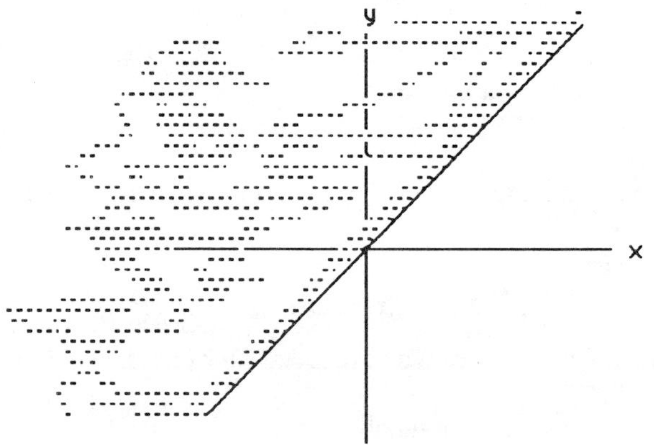

We have Dom(R) = Ran(R) = $\mathbb{R}$. •

Example

Let $A = \{1, \phi, \{\phi\}\}$ and $B = \{\phi, 2, \{1\}\}$. Then $R = \{(1,\phi), (\phi,\phi), (\{\phi\},\{1\})\}$ is a relation from A to B. Moreover, Dom(R) = $\{1, \phi, \{\phi\}\}$, and Ran(R) = $\{\phi, \{1\}\}$. •

Exercises for Section 3.1

1. Let $A = \{1, 2, 3, \phi\}$ and $B = \{\phi, \{1\}, 4, 5\}$. Let $R = \{(1,\phi), (1,\{1\}), (\phi,4)\}$. Is R a relation from A to B? Find Dom(R) and Ran(R).

2. Let $\mathbb{R}$ be the set of real numbers. Sketch the following relations from $\mathbb{R}$ to $\mathbb{R}$, and find the domains and ranges.
   a) $R = \{(x,y) \mid y = x^2\}$.
   b) $R = \{(x,y) \mid x = 2\}$.
   c) $R = \{(x,y) \mid y^2 = x\}$.
   d) $R = \{(x,y) \mid y \leq x^2\}$.

3. Let $A = \{1, 2, 3\}$ and $B = \{\phi, 2, \{1\}, \{\phi\}\}$. Write down two relations from A to B.

4. Let R be a relation from A to B. The relation R^{-1} from B to A defined by $R^{-1} = \{(r,s) \mid (s,r) \in R\}$ is called the inverse of the relation R. Find the inverses of the relations in the preceding exercises. What is the relation

between the domain and range of R and the domain and range of R^{-1}.

5. Let Φ be the empty set. Define $(a,b) = \{\{a,\Phi\},\{b,\{\Phi\}\}\}$. Prove that $(a,b) = (c,d)$ iff $a = c$ and $b = d$.

6. Define $(a,b) = \{a,\{a,b\}\}$. Prove that $(a,b) = (c,d)$ iff $a = c$ and $b = d$.

7. Let A, B, and C be sets. Prove that
   a) $(A \cup B) \times C = (A \times C) \cup (B \times C)$,
   b) $(A \cap B) \times C = (A \times C) \cap (B \times C)$, and
   c) $(A - B) \times C = (A \times C) - (B \times C)$.

8. In this exercise we define the **composition of relations.** Let A, B, C, and D be sets. Let R be a relation from A to B, and S be a relation from C to D. Then we define a relation from A to D as follows. For x in A and z in D, $(x,z) \in (S \circ R)$ iff there is a $y \in B \cap C$ such that $(x,y) \in R$, and $(y,z) \in S$.
   a) Let $A = \{1, 2, 3, 4\}$, $B = \{4, 5, 6\}$, and $C = D = \{2, 4, 6\}$. Let $R = \{(1,4), (1,6), (2,5)\}$, and $S = \{(2,4), (2,6), (4,6)\}$. Compute $S \circ R$.
   b) Let $A = B = C = D = \{1, 2, 3\}$. Show by example that we cannot expect $S \circ R = R \circ S$ in general.
   c) Let R and S be relations on A. Show that $(S \circ R)^{-1} = R^{-1} \circ S^{-1}$.
   d) As in b) show that we cannot expect $R^{-1} \circ S^{-1} = (R \circ S)^{-1}$.

3.2 Equivalence Relations

In this section we explore the properties of a very important type of relation called an equivalence relation. Also, we shall employ the notation xRy for $(x,y) \in R$.

Definition 3.2.1

Let X be a set. A relation R on X is an equivalence relation if and only if:
1) For all x in X, xRx.
2) For all x and y in X, if xRy then yRx.
3) For all x, y, and z in X, if xRy and yRz then xRz.

We call properties 1), 2) and 3) the reflexive, symmetric, and transitive laws, respectively. •

Examples

1) Let $X = \{1, 2, 3\}$. Define $R = \{(1,1), (2,2), (3,3), (1,2), (2,1)\}$, then R is an equivalence relation. That 1) and 2) hold is clear. That transitivity holds

is established by checking all cases.

2) Let $\mathbb{R}$ be the set of real numbers. Let xRy iff $x^2 = y^2$ for all x and y in $\mathbb{R}$.

3) Let xRy iff $\sin(x) = \sin(y)$ for all x and y in $\mathbb{R}$. •

It is very interesting to note that the sets of elements that are equivalent to given elements have, if they are distinct, empty intersections. Consider 1) in the above examples. We have, $\{x \mid xR1\} = \{1, 2\}$, $\{x \mid xR2\} = \{1, 2\}$ as before, and $\{x \mid xR3\} = \{3\}$. In 2) above we have $\{x \mid xRy\} = \{-y, y\}$. Sets that have an empty intersection are said to be disjoint. Thus, the sets $\{-y, y\}$ are disjoint or identical. Examples are $\{0\}$, $\{-1, 1\}$ and $\{-\pi, \pi\}$. The sets of equivalent elements in 3) are not so easy to describe. Picture the graph of the sine function. Draw a horizontal line that intersects the graph. The x coordinates of the points of intersection are equivalent. Perhaps the reader can also see that these sets are disjoint or identical. These observations motivate the following definition and theorem.

Definition 3.2.2

Let R be an equivalence relation on X. We define sets $[x]$ by

$$[x] = \{y \mid yRx\}.$$

We call $[x]$ the equivalence class of x. The family of all equivalence classes is denoted by X/R. That is, X/R is the set of sets defined by

$$X/R = \{[x] \mid x \in X\}. \bullet$$

It is standard to use the term "class" even though $[x]$ is a set.

Theorem 3.2.3

Let R be an equivalence relation on the set X. Then,

1) for all x in X, $[x] \neq \phi$, and

2) $\bigcup\{[x] \mid x \in X\} = X$, and

3) for all x and y in X, $[x] = [y]$, or $[x] \cap [y] = \phi$.

Proof:

Suppose R is an equivalence relation on the set X. Clearly $x \in [x]$, so $[x] \neq \phi$. Since $x \in [x]$ for all x in X, it follows that $\bigcup\{[x] \mid x \in X\} = X$.

Suppose $[x] \cap [y] \neq \phi$. Let $z \in [x] \cap [y]$. Then xRz and zRy, so xRy. Now let $c \in [x]$. Then cRx, and so cRy. Thus, $c \in [y]$. So, $[x] \subset [y]$. Similarly, $[y] \subset [x]$. So, $[x] = [y]$. •

A New Method of Proof

The reader may have noted that the proof of part 3) of Theorem 3.2.3 is a new method. The method proves a proposition of the form $P \Rightarrow (R \vee S)$ by adopting

P as the temporary premise as usual, and then taking $-S$ as an additional temporary premise and proving R from there. The idea behind it is simple; we prove $R \vee S$ is true by showing that it is not possible for R and S to both be false. The proof method shows that if S is false then R must be true. In the exercises the reader is asked to come up with a tautology that justifies the method.

Theorem 3.2.3 says that if R is an equivalence relation on X, then the equivalence classes, which are subsets of X, divide X into nonoverlapping regions much as Europe is divided into countries. This idea is important enough to warrant a special name.

Definition 3.2.4

A family $\mathfrak{F}$ of subsets of a set X is called a partition of X if and only if

1) for every A in $\mathfrak{F}$, $A \neq \phi$,
2) $\bigcup \{A \mid A \in \mathfrak{F}\} = X$, and
3) for all A and B in $\mathfrak{F}$, $A = B$ or $A \cap B = \phi$. •

An equivalence relation R on X creates a partition X/R of X. It is reasonable to ask if a partition $\mathfrak{F}$ of X creates an equivalence relation on X. The next theorem answers this question.

Theorem 3.2.5

Let $\mathfrak{F}$ be a partition of the set X. Define a relation R on X by xRy iff there exists an A in $\mathfrak{F}$ such that x and y are in A. Then R is an equivalence relation, and $X/R = \mathfrak{F}$.

Proof:
Let $x \in X$. Then $x \in A$ for some $A \in \mathfrak{F}$. (Why?) Therefore, xRx. Now suppose xRy, then $x, y \in A$ for some A. So, $y, x \in A$, and so yRx. Suppose xRy and yRz. Then x and y are in some A, and y and z are in some B. So y is in A and B, so $A = B$ (Why?). Therefore, x and z are in A. So, xRz. Finally, consider $[x]$. Let $x \in A$. Then $y \in [x]$ iff xRy, iff $y \in A$. Thus, $[x] = A$. Hence, $\mathfrak{F} = X/R$. •

Modular Arithmetic Revisited

Let $n \in \mathbf{N}$, $\mathbf{N}$ the natural numbers. We define a relation M_n on the integers $\mathbf{Z}$ as follows. We say xM_ny if and only if there is an integer k such that $x - y = kn$. In this case we say n divides $x - y$, we write $n \mid x - y$, and we say x is congruent to y modulo n.

Theorem 3.2.6

The relation M_n on **Z** is an equivalence relation.

Proof:
Clearly xM_nx. If xM_ny then $x - y = kn$ for some $k \in \mathbf{Z}$. Thus, $y - x = -kn$, and so yM_nx. Finally, if xM_ny and yM_nz then $x - y = k_1n$, and $y - z = k_2n$, so $x - z = (k_1 + k_2)n$ (Why?). Hence, xM_nz. •

Instead of $\mathbf{Z}/M_n$ we shall write $\mathbf{Z}_n$ for the family of equivalence classes of the equivalence relation M_n.

Example

Let $n = 4$. Let us examine $\mathbf{Z}_4$.

$[0] = \{\ldots, -8, -4, 0, 4, 8, \ldots\} = \{x \mid x = 0 + 4k, k \in \mathbf{Z}\}$
$[1] = \{\ldots, -7, -3, 1, 5, 9, \ldots\} = \{x \mid x = 1 + 4k, k \in \mathbf{Z}\}$
$[2] = \{\ldots, -6, -2, 2, 6, 10, \ldots\} = \{x \mid x = 2 + 4k, k \in \mathbf{Z}\}$
$[3] = \{\ldots, -5, -1, 3, 7, 11, \ldots\} = \{x \mid x = 3 + 4k, k \in \mathbf{Z}\}$

Note that [4], [5] etc are repetitions of the above classes. •

A little reflection shows that in general the distinct equivalence classes of M_n are [0], [1], [2], ... , $[n - 1]$. Indeed, given any integer x, by long division $x = kn + r$, and $0 \le r \le n - 1$, where r is unique.

The above example suggests another interesting property. Take a number from each of two equivalence classes, add them and see which class the sum falls in. Now repeat this with different numbers from the same two classes. The result still falls in the same class. For example, 5 + 6 = 11, and −7 + 10 = 3. This leads us to conjecture that the following theorem is true.

Theorem 3.2.7

Let M_n be the equivalence relation congruence modulo n on **Z**. For any x_1, x_2, y_1, and y_2 in **Z**, if $[x_1] = [x_2]$ and $[y_1] = [y_2]$ then $[x_1 + y_1] = [x_2 + y_2]$.

Proof:
Let $[x_1] = [x_2]$ and $[y_1] = [y_2]$, then $x_1 - x_2 = k_1n$ and $y_1 - y_2 = k_2n$. So, $(x_1 + y_1) - (x_2 + y_2) = (k_1 + k_2)n$, and so $[x_1 + y_1] = [x_2 + y_2]$ (Why?). •

In the exercises the reader will be asked to develop the same conjecture for multiplication. Accepting this, we make the following definition.

Definition 3.2.8

With M_n as above, and $[x]$ and $[y]$ in $\mathbf{Z}_n$, we define $[x] \oplus_n [y] = [x + y]$, and $[x] \otimes_n [y] = [x \times y]$. These operations are called addition and multiplication modulo n. •

The principles involved in arriving at this definition are so important and subtle that we devote the closing portion of this section to them.

Well-Definition

In defining the modular sum $[x] \oplus_n [y]$ of two classes to be the class of the sum of representatives of the classes we expose ourselves to the following danger of contradiction. Since we can choose any one of infinitely many integers to represent $[x]$ and $[y]$, it could have happened that the resulting sums would not be in the same class. In that case the contradiction would be to conclude that two unequal sets were equal. Of course Theorem 3.2.7 proves that we cannot fall into this contradiction. Let us consider an example where we can.

Let $R = \{(x,y) \mid x^2 = y^2\} \subset \mathbb{R} \times \mathbb{R}$, then R is an equivalence relation. Now define $[x] \oplus [y] = [x + y]$. Thus $[1] \oplus [2] = [3]$, but $[1] = [-1]$, and so $[1] \oplus [2] = [1]$. So $[1] = [3]$, and $[1] \neq [3]$. We are in trouble!

The moral is that whenever we use representatives of sets to define operations on these sets we must always show that such contradictions cannot result. Doing so is called showing that the definition is well-defined.

Exercises for Section 3.2

1. Give examples of relations that have just two of the three properties of an equivalence relation.

2. Let X be a set. Consider "=" as a relation on X. Show that "=" is an equivalence relation. What are its equivalence classes?

3. Give three examples of equivalence relations on $\mathbb{R}$ that are not examples in this book. Explain why your examples are correct. Describe the equivalence classes.

4. Let $\mathcal{F} = \{\{0\}, \mathbb{R} - \{0\}\}$. Here $\mathcal{F} \subset \mathcal{P}(\mathbb{R})$. Describe the equivalence relation this partition defines.

5. Let $X = \{1, 2, 3, 4\}$. Let $\mathcal{F} = \{\{1, 3\}, \{2, 4\}\}$. $\mathcal{F}$ defines an equivalence relation R on X. Exhibit it as a set of ordered pairs, and show that its set of equivalence classes is $\mathcal{F}$.

6. If R is an equivalence relation on X prove that R^{-1} is also.

7. Find a tautology that justifies the new method of proof that was introduced in this section.

8. Following the example in the text:
   a) Exhibit the equivalence classes of M_5 on **Z**.
   b) Give some examples that suggest that multiplication modulo 5 is well-defined.

9. Prove that multiplication modulo n is well-defined.

10. Prove that for every x, y, and z in **Z**,
$$[x] \otimes_n \{[y] \oplus_n [z]\} = \{[x] \otimes_n [y]\} \oplus_n \{[x] \otimes_n [z]\}.$$

3.3 Functions

In this section we define the concept of a function in terms of the concept of a relation.

Definition 3.3.1

Let A and B be sets. A function from A to B is a relation f from A to B such that:
1) $\mathrm{Dom}(f) = A$, and
2) for every x in A and y_1 and y_2 in B, if $(x,y_1) \in f$ and $(x,y_2) \in f$ then $y_1 = y_2$. •

Naturally enough, we call A the domain of the function. The reader should note that B is not in general the range of f. We call B the co-domain of the function. The standard notation for a function f from A to B is, $f: A \to B$. We read this symbol as "f is a function from A to B", or "f is a map or mapping from A to B". In order for a function to be defined we shall insist that the domain A, the co-domain B, and the relation f from A to B be given.

For relations it is quite possible for some x in A to be related to several y in B. But for a function $f: A \to B$, for each x in A there is exactly one y in B such that f relates x to y. This feature of functions allows us to use the familiar notation $f(x)$ for y. So, $f(x)$ is the element in B to which f associates x.

Examples

1) Let $A = \{1, 2, 3\}$ and $B = \{4, 5\}$. Let $f = \{(1,4), (2,4), (3,4)\}$. Then,

$f : A \to B$ is a function. Note that Ran(f) = $\{4\} \neq B$. The function notation gives $f(1) = 4$ for example.

2) Let $A = B = \mathbb{R}$. Define $f(x) = x^2$. Then $f : A \to B$ is a function. In this case $f(2) = 4$, and $f(-3) = 9$ for example. Note how we have used $f(x) = x^2$ in place of $f = \{(x,y) \mid y = x^2\}$.

3) Let $A = B = \mathbb{R}$. Let, $f = \{(x,y) \mid \text{if } x \in [0,1] \text{ then } y = 1, \text{ and if } x \notin [0,1] \text{ then } y = 0\}$.

Clearly this defines a function from $\mathbb{R}$ to $\mathbb{R}$. But it would have been easier to define f as follows. Let $f(x) = 1$ if $x \in [0,1]$, and $f(x) = 0$ otherwise. •

In this next example we illustrate the need to take care to see that a function is well-defined when equivalence classes are involved.

Example

Consider the equation $f([x]_3) = [x]_6$ where $[x]_3 \in \mathbf{Z}_3$ and $[x]_6 \in \mathbf{Z}_6$. The question is whether this equation defines a function $f : \mathbf{Z}_3 \to \mathbf{Z}_6$. It does not. The function is not well defined since $[1]_3 = [4]_3$, but $[1]_6 \neq [4]_6$.

On the other hand, $g : \mathbf{Z}_6 \to \mathbf{Z}_3$ defined by $g([x]_6) = [x]_3$ is well defined. Indeed, if $[x]_6 = [y]_6$ then $6 \mid x - y$, so $3 \mid x - y$. Thus, $[x]_3 = [y]_3$. •

The next example is intended to dispel the misconception that all functions are formulas.

Example

Define $f : \mathrm{N} \to \mathrm{N}$ by $f(n)$ is the nth decimal place in the decimal expansion of π. So $f(1) = 1, f(2) = 4, f(3) = 1$, etc. But $f(1000) = ?$ •

The reader who has suffered through the usual college calculus course can hardly be blamed for thinking that the notion of a function means an equation. To be sure we use equations to define functions very often. The next example shows that the same equation can define different functions.

Example

Define $f : \mathbb{R} \to \mathbb{R}$ by $f(x) = x^2$. Define $g : [0,10] \to \mathbb{R}$ by $g(x) = x^2$. The equations are the same, but these functions are not. The domains are not equal.

Define $h : \mathbb{R} \to [0,\infty)$ by $h(x) = x^2$. Then $f : \mathbb{R} \to \mathbb{R}$ and $h : \mathbb{R} \to [0,\infty)$ are not the same function since they do not have the same co-domains. •

The reader should be warned that some authors would regard the last pair of functions as equal.

When are functions equal? Consider the functions $f : A \to B$ and $g : C \to D$. These functions are equal if and only if $A = C$, $B = D$, and $f(x) = g(x)$ for all x in A (or C). This latter condition is clearly necessary since, as relations, $f = g$ iff $f(x) = g(x)$ for all x in A.

That A should equal C and $f = g$ as relations is clearly required in anyone's concept of the equality of functions, but why should we require that $B = D$? The reason is that a function $f : A \to B$ not only relates the elements of A to some of the elements of B, but f puts A as a whole in relation to B as a whole in a way that can prove to be important in modern mathematics. In addition, if functions are equal we want to be able to make exactly the same statements about each. This would not be so for the functions $f : \mathbb{R} \to \mathbb{R}$ and $h : \mathbb{R} \to [0,\infty)$ defined in the last example.

While functions are not identical with equations, equations are used quite often to define functions. For example, one will often see expressions such as "consider the function $y = 1/(x - 1)$". In order for such an expression to define a function for us we must introduce a convention for determining the domain and co-domain in such cases. The convention applied to this case is to consider the relation

$$f = \{(x,y) \mid y = 1/(x - 1)\} \subset \mathbb{R} \times \mathbb{R},$$

and then take the domain to be Dom(f) and the co-domain to be $\mathbb{R}$. With regard to the domain, we are observing the usual convention of taking the domain to be the largest subset of $\mathbb{R}$ for which the equation makes sense.

Now that we have the concept of a function clearly in mind, we are in a position to show that the familiar operations of algebra are functions.

Binary Operations

The operations of addition (+) and multiplication (×) on $\mathbb{R}$ are functions. For example, given a pair of numbers $(x,y) \in \mathbb{R} \times \mathbb{R}$, + associates the number $x + y$ with the pair. We have a function $+ : \mathbb{R} \times \mathbb{R} \to \mathbb{R}$. Thus, we write $+((x,y)) = x + y$. This motivates the following definition.

Definition 3.3.2

A binary operation on a set A is a function of the form $b : A \times A \to A$. •

Example

Consider the function $m : \mathbb{R} \times \mathbb{R} \to \mathbb{R}$ defined by $m((x,y)) = (x + y)/2$. So, for example $m((2,6)) = 4$. The function m is a binary operation. Of course, it is the operation of taking the mean or average of x and y. •

Example

The operations of addition and multiplication modulo n are binary operations because $\oplus_n : \mathbf{Z}_n \times \mathbf{Z}_n \to \mathbf{Z}_n$, and $\otimes_n : \mathbf{Z}_n \times \mathbf{Z}_n \to \mathbf{Z}_n$ defined by $\oplus_n(([x],[y])) = [x] \oplus_n [y]$ and $\otimes_n(([x],[y])) = [x] \otimes_n [y]$ are well-defined. •

Exercises for Section 3.3

1. Let $A = \{1, 2, 3\}$ and $B = \{a, b, c, d\}$. Which of the following relations f make $f: A \to B$ into a function from A to B? Give reasons.
   a) $f = \{(1,a), (2,a), (3,a)\}$. b) $f = \{(1,a), (1,b), (2,c), (3,d)\}$.
   c) $f = \{(5,a), (2,b), (3,c)\}$. d) $f = \{(1,c), (2,d), (3,\pi)\}$.

2. Let $A = \{1, 2, 3\}$ and $B = \{a, b, c, d\}$. What are the ranges of the following functions from A to B?
   a) $f = \{(1,a), (2,a), (3,a)\}$. b) $f = \{(1,b), (2,c), (3,c)\}$.

3. Define $f: \mathbb{R} \to \mathbb{R}$ by $f(x) = 1/(x^2 + 1)$. What is f as a set of ordered pairs? What is the range of $f: \mathbb{R} \to \mathbb{R}$, and what is its co-domain?

4. Define $f: \mathbb{R} \to \mathbb{R}$ by $f(x) = \sqrt{x^2}$, and $g: \mathbb{R} \to \mathbb{R}$ by $g(x) = |x|$ (absolute x). Are these functions equal? Explain !

5. Define $f: \mathbb{R} \to \mathbb{R}$ by $f(x) = \sqrt{x^2}$, and $g: \mathbb{R} \to [0,\infty)$ by $g(x) = |x|$. Are these functions equal? Explain !

6. Find the domains, ranges and co-domains of the functions defined by the following equations.
   a) $y = 2x + 1$. b) $y = \sqrt{1 - x^2}$. c) $y = 1/(x^2 - 1)$.
   d) $y = \sqrt{7 - x} + \sqrt{x - 1}$.

7. Let $A = \{1, 2, 3\}$ and $B = \{a, b, c, d\}$. How many different functions are there from A to B? Explain ! Generalize !

8. Define two binary operations on $\mathbb{R}$ other than the examples given in the text.

9. Express the associative law of addition, $x + (y + z) = (x + y) + z$, in terms of the notation for + as a function. Which do you prefer?

3.4 Composition of Functions

If one thing is related to a second and the second is related to a third, then this relates the first to the third. This idea was defined formally in Exercise 7 of Section 1 of this chapter. We repeat it here for completeness.

Definition 3.4.1

Let A, B, C, and D be sets. Let R be a relation from A to B, and S be a relation from C to D. Then we define a relation from A to D as follows. For x in A and z in D, $(x,z) \in (S \circ R)$ iff there is a $y \in B \cap C$ such that $(x,y) \in R$, and $(y,z) \in S$. Then $S \circ R$ is a relation from A to D. It is called the composition of R and S. •

The most important type of composition of relations occurs when the relations are functions, and it is on this situation that we shall focus. Let $f : A \rightarrow B$ and $g : C \rightarrow D$ be functions. Then Definition 3.4.1 defines the composition $g \circ f$ as a relation from A to D. However, without any restrictions it is not likely that $g \circ f$ will be a functon from A to D. To ensure that $g \circ f$ is a function from A to D we require that $\text{Ran}(f) \subset C$. We make the following formal definition.

Definition 3.4.2

Let $f : A \rightarrow B$ and $g : C \rightarrow D$ be functions. Let $\text{Ran}(f) \subset C$. Then we define $g \circ f : A \rightarrow D$ by $g \circ f(x) = g(f(x))$ for all x in A. We call $g \circ f : A \rightarrow D$ the composition of f with g. •

Theorem 3.4.3

Let $f : A \rightarrow B$ and $g : C \rightarrow D$ be functions. Let $\text{Ran}(f) \subset C$. Then $g \circ f$ is a function.

Proof:
Let (x,z_1) and (x,z_2) be in $g \circ f$. Then we have y_1 and y_2 in C such that $g(y_1) = z_1$, and $g(y_2) = z_2$. But $y_1 = f(x) = y_2$, and so $z_1 = z_2$. Clearly, $\text{Dom}(f) = A$. Therefore, $g \circ f$ is a function. •

Notice that the relation $g \circ f$ defined here agrees with that produced by the composition of relations defined above.

The reader will find it useful to picture functions and compositions of functions as shown below.

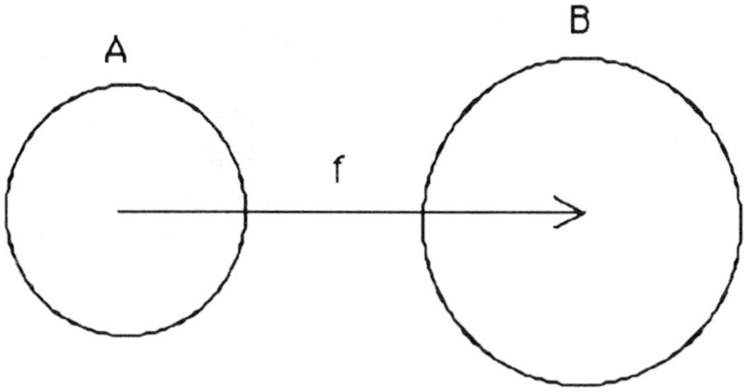

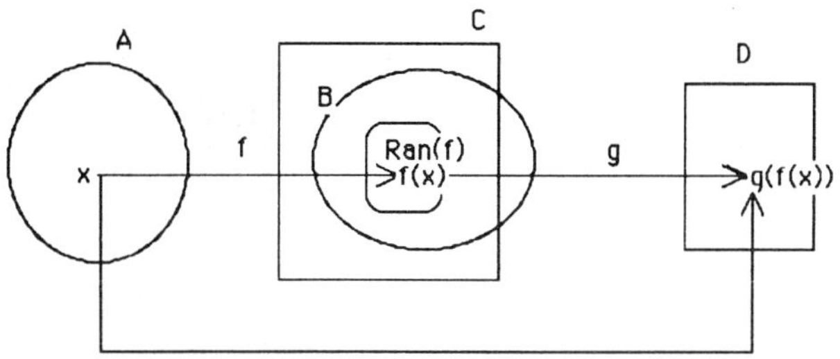

$g\circ f$

Example

Let $f: \mathbb{R} \to \mathbb{R}$ be defined by $f(x) = x^2$, and $g: \mathbb{R} \to \mathbb{R}$ be defined by $g(x) = x + 4$. Then $g\circ f: \mathbb{R} \to \mathbb{R}$ is defined by

$$g\circ f(x) = g(f(x)) = g(x^2) = x^2 + 4.$$

Note that $f\circ g(x) = (x + 4)^2$, so $g\circ f \neq f\circ g$. •

Example

Let $A = \{1, 2, 3\}$, $B = \{a, b, c, d\}$, and $C = \{5, 6, 7\}$. Define $f: A \to B$ by $f = \{(1,a), (2,a), (3,b)\}$ and $g: B \to C$ by $g = \{(a,5), (b,6), (c,5), (d,7)\}$. Then, $g\circ f: A \to C$ is given by $g\circ f = \{(1,5), (2,5), (3,6)\}$. •

When two compositions can be performed in succession the operation is associative as we now prove.

Theorem 3.4.4

Let $f: A \to B$, $g: C \to D$, and $h: E \to F$. Suppose $\text{Ran}(f) \subset C$, and $\text{Ran}(g) \subset E$. Then $[h\circ(g\circ f): A \to F] = [(h\circ g)\circ f: A \to F]$.

Proof:
Let $x \in A$. Then $h\circ(g\circ f)(x) = h((g\circ f)(x)) = h(g(f(x))) = (h\circ g)(f(x)) = (h\circ g)\circ f(x)$. Thus, $h\circ(g\circ f) = (h\circ g)\circ f$, and the domains and co-domains are the same. So the functions are equal. •

When it is clear that we are talking about the equality of functions we shall write $f = g$ in place of $[f : A \to B] = [g : C \to D]$.

We conclude this section with a definition that will be useful in the next section.

Definition 3.4.5

Let A be a set. We define $I_A : A \to A$ by $I_A(x) = x$. We call I_A the identity function on A. •

Note that if $f : A \to B$ is a function then $f\circ I_A = f$, and $I_B\circ f = f$.

Exercises for Section 3.4

1. Let $A = \{1, 2, 3\}$, $B = \{2, 5, 6, 7\}$, $C = \{1, 2, 3, 5, 6, 7\}$, and $D = \{a, b, c, d\}$. Let $f : A \to B$ and $g : C \to D$ be defined by $f = \{(1,2), (2,6), (3,7)\}$ and $g = \{(1,a), (2,a), (3,d), (5,c), (6,b), (7,d)\}$. Compute $g\circ f : A \to D$.

2. Let $f : \mathbb{R} \to \mathbb{R}$ and $g : \mathbb{R} \to \mathbb{R}$ be given by the formulas below. Complete $g\circ f(x) =$, and $f\circ g(x) =$, and find the range of each composition.
a) $f(x) = x^3 + 1$, $g(x) = x^2 + 2x$. b) $f(x) = x^2$, $g(x) = 3$.

3. Can the following functions be composed in either order? Explain! Let $f : \mathbb{R} \to \mathbb{R}$ and $g : [-1,1] \to \mathbb{R}$ be defined by $f(x) = x + 2$, and $g(x) = \sqrt{1 - x^2}$.

4. Let $f : \mathbb{R} \to \mathbb{R}$ be a function with the curious property that for all x and y in $\mathbb{R}$, $f(x + y) = f(x) + f(y)$. Prove that the composition of two such functions has the same curious property. Give two examples of such functions.

5. Let $f : \mathbb{R} \to \mathbb{R}$ be defined by $f(x) = mx + b$. We call such mappings of $\mathbb{R}$ affine. Prove that the composition of affine maps is affine.

3.5 Inverses, Injections, Surjections, and Bijections

In this section we describe the concepts of one-sided and two-sided inverses of a function. We begin with the definition of the inverse of a relation.

Definition 3.5.1

Let R be a relation from A to B. We define R^{-1} by

$$R^{-1} = \{(x,y) \mid (y,x) \in R\}.$$

Then R^{-1} is a relation from B to A, and we call it the inverse of R. •

Observe that $\text{Dom}(R^{-1}) = \text{Ran}(R)$ and $\text{Ran}(R^{-1}) = \text{Dom}(R)$.

Let $f : A \to B$ be a function, then we can ask what is necessary for f^{-1} to be a function from B to A. We must have $\text{Dom}(f^{-1}) = B$, so we need $\text{Ran}(f) = B$. This condition is so important that we give it a special name.

Definition 3.5.2

We say $f : A \to B$ is a surjection if and only if $\text{Ran}(f) = B$. In this case we often say that f is "onto B", or simply "onto". •

For f^{-1} to be a function we shall also need f^{-1} to satisfy 2) of Definition 3.3.1. In this connection we have the following theorem.

Theorem 3.5.3

The relation f^{-1} from B to A satisfies 2) of Definition 3.3.1 iff for every x_1 and x_2 in A, if $f(x_1) = f(x_2)$ then $x_1 = x_2$.

Proof:
Suppose f^{-1} satisfies 2) of Definition 3.3.1. Suppose $f(x_1) = f(x_2) = y$. Then $(x_1,y) \in f$ and $(x_2,y) \in f$. So, $(y,x_1) \in f^{-1}$ and $(y,x_2) \in f^{-1}$. Therefore, $x_1 = x_2$ by 2).

Conversely, suppose that for every x_1 and x_2 in A, if $f(x_1) = f(x_2)$ then $x_1 = x_2$. Suppose that $(y,x_1) \in f^{-1}$ and $(y,x_2) \in f^{-1}$. Then, $(x_1,y) \in f$ and $(x_2,y) \in f$. So, $f(x_1) = f(x_2)$, and so $x_1 = x_2$. •

The property described in the theorem is again important enough to be named.

Definition 3.5.4

Let $f : A \to B$ be a function such that for every x_1 and x_2 in A, if $f(x_1) = f(x_2)$ then $x_1 = x_2$. Then we say that f is an injection, or that f is one to one. •

Corollary 3.5.5

The function $f : A \to B$ has an inverse iff f is an injection and a surjection.

Recall that a corollary is a simple consequence of preceding results.

Definition 3.5.6

A function which is both an injection and a surjection is called a bijection. •

We now give some examples of the concepts just developed. The examples include proofs which use properties of the real numbers. As usual we assume any needed properties of the reals.

Example

Let $f : \mathbb{R} \to [0,\infty)$ be defined by $f(x) = x^2$. Then f is a surjection (onto).

Proof:
Since we must prove that $\text{Ran}(f) = B = [0,\infty)$, and since it is always the case that $\text{Ran}(f) \subset B$, we prove that $B \subset \text{Ran}(f)$. Let $y \in [0,\infty)$, then $y \geq 0$, so $\sqrt{y} \in \mathbb{R}$. So, $f(\sqrt{y}) = (\sqrt{y})^2 = y$, and so $y \in \text{Ran}(f)$. •

Example

Let $f : [0,\infty) \to \mathbb{R}$ be defined by $f(x) = x^2$. Then f is an injection (one to one).

Proof:
Let x_1 and x_2 belong to $\mathbb{R}$. Let $f(x_1) = f(x_2)$, so $x_1{}^2 = x_2{}^2$. Therefore, $\sqrt{x_1{}^2} = \sqrt{x_2{}^2}$, and so $|x_1| = |x_2|$. But $x_1 \geq 0$ and $x_2 \geq 0$, so $x_1 = x_2$. •

Example

Let $f : \mathbb{R} \to \mathbb{R}$ be defined by $f(x) = x^3 + 2$. Then, f is a bijection.

Proof:
Let x_1 and x_2 be in $\mathbb{R}$. Suppose $f(x_1) = f(x_2)$, then $x_1{}^3 + 2 = x_2{}^3 + 2$, so $x_1{}^3 = x_2{}^3$, and so $x_1 = x_2$. Thus, f is an injection.

Let $y \in \mathbb{R}$. We need to find an x in $\mathbb{R}$ such that $f(x) = y$. Suppose $y = x^3 + 2$, then $x = \sqrt[3]{y-2}$. So given y in $\mathbb{R}$ there is an x in Dom(f), namely $\sqrt[3]{y-2}$, such that $f(x) = f(\sqrt[3]{y-2}) = (\sqrt[3]{y-2})^3 + 2 = y$. Thus, $y \in$ Ran(f), and so f is a surjection. •

The second paragraph of the above proof deserves comment. The sentence "Suppose $y = x^3 + 2$, then $x = \sqrt[3]{y-2}$." seems to assume what we are trying to prove, namely, that there is an x such that $y = x^3 + 2$. This is, in fact, true. The sentence quoted does not belong in a strictly formal proof. It is an aside to the reader explaining where the quantity $\sqrt[3]{y-2}$ came from. In this proof we must demonstrate the truth of the proposition:

"For every y in $\mathbb{R}$ there is an x in $\mathbb{R}$ such that $f(x) = y$".

One way to do this is to exhibit a correspondence $y \to x_y$ such that $f(x_y) = y$. Our quoted sentence shows how the correspondence $y \to \sqrt[3]{y-2}$ was discoverd, then the proof goes on to show that $f(x_y) = y$ is true.

Such asides or investigative digressions are commonplace in informal proofs. They act as explanatory material, but often they are not strictly part of the proof and could be discarded without effecting the validity of the proof. The reader must be on the look-out for such digressions that go against the logical direction of a proof.

Here are two more examples of proofs of the injective and surjective properties.

Example

Let $f : (0,1) \to \mathbb{R}$ be defined by $f(x) = 1/x$. Then, f is an injection.

Proof:
Choose x_1 and x_2 in (0,1). Let $f(x_1) = f(x_2)$, then $1/x_1 = 1/x_2$. Therefore, $x_1 = x_2$, and so f is one to one. •

Example

Let $f : (0,1) \to (1,\infty)$ be defined by $f(x) = 1/x$. Then, f is a surjection.

Proof:
We must show that $\text{Ran}(f) = (1,\infty)$. Of course, $\text{Ran}(f) \subset (1,\infty)$. Choose y in $(1,\infty)$, then $y > 1$, and $1/y < 1$. Set $x = 1/y$, then $f(x) = f(1/y) = y$. So, y is in $\text{Ran}(f)$. So, $(1,\infty) \subset \text{Ran}(f)$. Thus, $\text{Ran}(f) = (1,\infty)$, and f is onto. •

Note that if $f : A \to B$ has an inverse $f^{-1} : B \to A$ then $f \circ f^{-1} = I_B$, and $f^{-1} \circ f = I_A$. This observation motivates the following theorem.

Theorem 3.5.7

Let $f : A \to B$. Then f has an inverse iff there exists a function $g : B \to A$ such that $g \circ f = I_A$ and $f \circ g = I_B$.

Proof:
If f has an inverse f^{-1} then as we have said $f^{-1} \circ f = I_A$ and $f \circ f^{-1} = I_B$.

Conversely, suppose there exists a function $g : B \to A$ such that $g \circ f = I_A$ and $f \circ g = I_B$. We shall show that f is a bijection. Suppose $y \in B$. Set $x = g(y)$. Then $f(x) = f(g(y)) = f \circ g(y) = I_B(y) = y$. Therefore, f is onto.

Now suppose $f(x_1) = f(x_2)$, then $g(f(x_1)) = g(f(x_2))$. So, $g \circ f(x_1) = g \circ f(x_2)$. Thus, $x_1 = x_2$ (Why?). Therefore, f is one to one. •

This theorem motivates the following definition.

Definition 3.5.8

Let $f : A \to B$ and $g : B \to A$. If $g \circ f = I_A$ we say g is a left inverse of f, and if $f \circ g = I_B$ we say g is a right inverse of f. If both conditions hold we say g is a two-sided inverse of f. •

We leave the reader to prove that if g is a two-sided inverse of f then $g = f^{-1}$.

We now characterize left and right inverses.

Theorem 3.5.9

The function $f : A \to B$ has a left inverse iff $f : A \to B$ is an injection.

Proof:
Suppose f has a left inverse $g : B \to A$. Now suppose $f(x_1) = f(x_2)$.
Then, $g(f(x_1)) = g(f(x_2))$, and so $x_1 = x_2$. (Why?)

Suppose f is an injection. Let $a \in A$. For $y \in \text{Ran}(f)$ define $g(y) = x$ where x is the unique element of A such that $f(x) = y$. For $y \in B$ and $y \notin \text{Ran}(f)$ set $g(y) = a$. Then for any x in A, $g \circ f(x) = g(f(x)) = x$. So, $g \circ f = I_A$, and so g is a left inverse of f. •

A diagram will help illuminate the above proof.

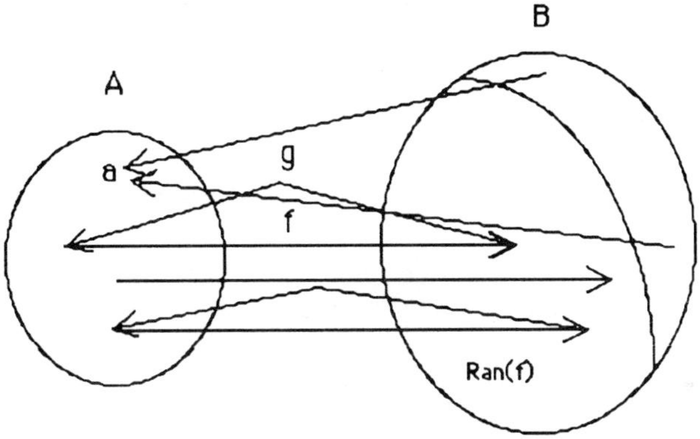

Each element in the range of f is sent back to where it came from, and those elements not in the range are all sent to some arbitrary element a of A.

Theorem 3.5.10

The function $f : A \to B$ has a right inverse iff $f : A \to B$ is a surjection.

Proof:
Suppose $f : A \to B$ has a right inverse $g : B \to A$. Then, $f \circ g = I_B$. Let $y \in B$. Then, $g(y) \in A$, and $f(g(y)) = f \circ g(y) = y$. So, $y \in \text{Ran}(f)$, and so f is a surjection.

Suppose f is a surjection. Let $y \in B$. Let $S_y = \{x \mid f(x) = y\}$. Note that because f is onto $S_y \neq \phi$. Choose x_y in S_y, and define $g(y) = x_y$. Then, $g : B \to A$, and $f \circ g(y) = f(g(y)) = f(x_y) = y = I_B(y)$. So, $f \circ g = I_B$. •

Again, a diagram will help in understanding the second part of this proof.

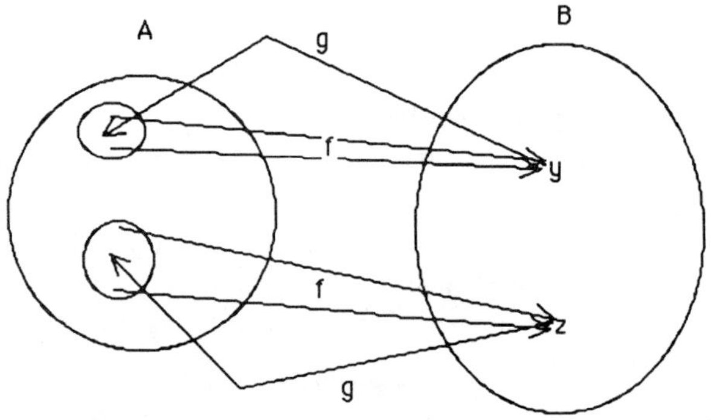

The patches in A represent the elements in A that f relates to y and z respectively. The function g relates y and z to one element only in each of these patches. Note that the patches are necessarilly disjoint.

Exercises for Section 3.5

1. Prove that if $g : B \to A$ is a two-sided inverse of $f : A \to B$ then $g = f^{-1}$.

2. Prove, using Theorem 3.5.9 that the functions $f : (0,\infty) \to \mathbf{R}$ defined by the following formulas have left inverses. Define one left inverse for each.
   a) $f(x) = x^2 + 1$. b) $f(x) = 3x + 2$. c) $f(x) = \dfrac{1}{x^3 + 2}$.

3. Prove that the following functions $f : \mathbb{R} \to \mathbb{R}$ are surjections.
   a) $f(x) = 2x - 7$. b) $f(x) = 1/(x - 1), f(1) = 0$. c) $f(x) = x^5 + 1$.

4. Prove that $f : \mathbb{R} \to \mathbb{R}$ defined by $f(x) = x^3 - 3x$ has a right inverse. Draw the graph of one such right inverse. Hint: You may want to use a little calculus. Does f have a left inverse? Why?

5. We say a function $f : \mathbb{R} \to \mathbb{R}$ is increasing if and only if for all x_1 and x_2 in $\mathbb{R}$, if $x_1 < x_2$ then $f(x_1) < f(x_2)$. Prove that if f is increasing then f is an injection. Is the converse true?

6. Let $f : A \to B$, and $g : B \to C$. Prove a) if f and g are injections then $g \circ f$ is an injection, and b) if f and g are surjections then $g \circ f$ is a surjection.

7. Let $f : \mathbb{R} \to \mathbb{R}$ be defined by $f(x) = 1/x$ if $x \neq 0$, and $f(0) = 0$. Prove that f has an inverse and find it.

8. Let $A = \{1, 2, 3, 4, 5\}$ and $B = \{a, b, c\}$. Let,

$$f = \{(1,a), (2,a), (3,b), (4,c), (5,c)\}.$$

Find a right inverse for f. Prove that f cannot have a left inverse.

9. Construct a finite example, like 8) above, of a function f that has a left inverse but no right inverse.

10. Give examples of functions f and g with the following properties.
   a) f is one to one, g is not, but $g \circ f$ is one to one.
   b) g is onto, f is not, but $g \circ f$ is onto.

3.6 The Axiom of Choice (Optional)

This section may be omitted without loss of continuity until Chapter 6. However, the topic covered in this section occupies a central place in modern mathematics, and so the reader should, at least, give this section a quick look over.

In the proof of the second part of Theorem 3.5.10 we defined the function g by choosing an element x_y from each of the sets $S_y = \{x \mid f(x) = y\}$. The process amounted to this. Given a family of nonempty sets $\mathfrak{F}$, we assumed there was a function $h : \mathfrak{F} \to \cup\mathfrak{F}$ with the property that for each $A \in \mathfrak{F}$, $h(A) \in A$. This function h formally describes what we mean by choosing an element out of each of the sets A in the family $\mathfrak{F}$. Such functions are called choice functions. The question is: "what guarantees the existence of choice functions?". It has been shown that the existence of choice functions cannot be deduced from our axioms of set theory that we laid down in Chapter 2. In view of this, we are in need of a new axiom to justify such choices.

The Axiom of Choice

Given any nonempty family of nonempty sets $\mathfrak{F}$, there exists a function $h : \mathfrak{F} \to \cup\mathfrak{F}$ with the property that for each $A \in \mathfrak{F}$, $h(A) \in A$. •

The Axiom of Choice is equivalent to Zorn's Lemma and to other very useful principles of set theory. For further information on these concepts we refer the reader to Chapter 6 and the reference at the end of this chapter.

3.7 Image and Inverse Image Maps

Let $f: A \to B$. Then f relates the elements of a subset of A to those of a subset of B. In addition, a subset of B is related to that subset of A composed of all those elements in A that f relates to elements in the subset of B. We formally define these maps.

Definition 3.7.1

Let $f: A \to B$. Define $f_* : \mathcal{P}(A) \to \mathcal{P}(B)$ by $f_*(S) = \{f(x) \mid x \in S\}$, where $S \subset A$. Define $f^* : \mathcal{P}(B) \to \mathcal{P}(A)$ by $f^*(T) = \{x \mid x \in A \text{ and } f(x) \in T\}$, where $T \subset B$. We call f_* the image map and f^* the inverse image map. Also, we call $f_*(S)$ the image of S under f and $f^*(T)$ the inverse image of T under f. •

Example

Let $A = \{1, 2, 3, 4\}$ and $B = \{5, 6, 7\}$. Let $f(1) = 5, f(2) = 5, f(3) = 6$, and $f(4) = 6$. Then, $f_*(\{1, 2\}) = \{5\}$, and $f_*(\{2, 4\}) = \{5, 6\}$. Also, $f^*(\{6\}) = \{3,4\}$, and $f^*(B) = A$. •

Example

Let $f: \mathbb{R} \to \mathbb{R}$ be defined by $f(x) = x^2$. Then $f_*([0,2)) = [0,4)$, and $f^*([1,2]) = [-\sqrt{2},-1] \cup [1,\sqrt{2}]$. •

The following diagrams should help in understanding these concepts.

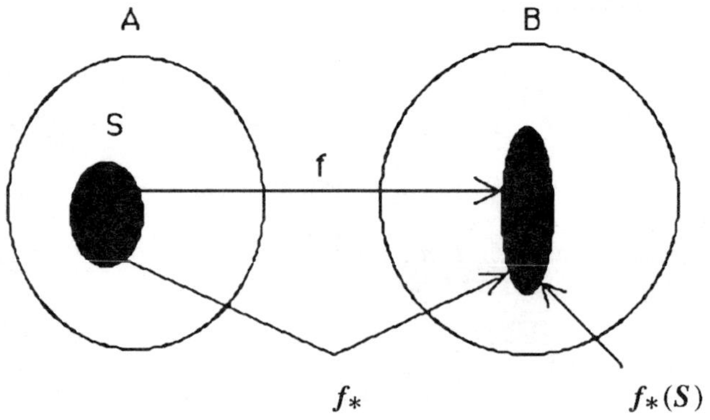

The corresponding figure for the inverse image is :

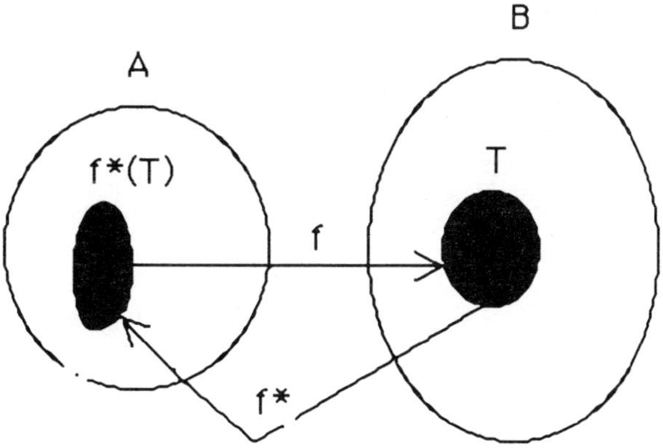

We now turn to proving some theorems about the image and inverse image maps.

Theorem 3.7.2

Let $f : A \to B$, and let $\mathfrak{F}$ be a family of subsets of A and $\mathfrak{G}$ a family of subsets of B. Then,

a) $f_*(\bigcap \mathfrak{F}) \subset \bigcap \{f_*(S) \mid S \in \mathfrak{F}\}$, and

b) $f_*(\bigcup \mathfrak{F}) = \bigcup \{f_*(S) \mid S \in \mathfrak{F}\}$, and

c) $f^*(\bigcap \mathfrak{G}) = \bigcap \{f^*(T) \mid T \in \mathfrak{G}\}$, and

d) $f^*(\bigcup \mathfrak{G}) = \bigcup \{f^*(T) \mid T \in \mathfrak{G}\}$.

Proof: (of a)

Let $y \in f_*(\bigcap \mathfrak{F})$. Then there is an x in $\bigcap \mathfrak{F}$ such that $f(x) = y$. But, $x \in S$ for all $S \in \mathfrak{F}$. So, $f(x) \in f_*(S)$ for all $S \in \mathfrak{F}$. So, $y \in \bigcap \{f_*(S) \mid S \in \mathfrak{F}\}$. •

The proofs of b), c), and d) are left as exercises.

Theorem 3.7.3

Let $f : A \to B$. Let $T \subset B$. Then $f^*(B - T) = A - f^*(T)$.

Proof:

Let $x \in f^*(B - T)$ then $f(x) \in B - T$. So, $f(x) \notin T$, so $x \notin f^*(T)$. So, $x \in A - f^*(T)$.

Conversely, if $x \in A - f^*(T)$, then $x \notin f^*(T)$, and so $f(x) \notin T$. Thus,

$f(x) \in B - T$, and so $x \in f^*(B - T)$. •

Clearly, f_* and f^* can be composed. We now discuss the relations between $f^*(f_*(S))$ and S, and between $f_*(f^*(T))$ and T. We shall make several assertions that are left to the reader as exercises.

It is obvious that $S \subset f^*(f_*(S))$, and $f_*(f^*(T)) \subset T$. We may then ask when, if ever, will equality hold. The reason that $S \subset f^*(f_*(S))$ can be a proper subset relationship is that f may not be one to one. If this is the case the image of a singleton will be a singleton, but its inverse image may not be. We contend that we will have equality if and only if f is one to one.

On the other hand, equality fails to hold in $f_*(f^*(T)) \subset T$ when T is not a subset of the range of f. Indeed, this is obvious since $f_*(f^*(T)) \subset \text{Ran}(f)$. We conjecture that we have equality if and only if f is a surjection.

Exercises for Section 3.7

1. Let $A = \{1, 2, 3, 4\}$, and $B = \{a, b, c, d\}$. Define $f : A \to B$ by $f(1) = c$, $f(2) = b$, $f(3) = c$, and $f(4) = a$. Compute f_* and f^*.

2. Prove parts b), c), and d) of Theorem 3.7.2.

3. When is $f_* \circ f^* = I_{\mathcal{P}(B)}$? Prove your conjecture.

4. When is $f^* \circ f_* = I_{\mathcal{P}(A)}$? Prove your conjecture.

5. Find a counter example to show that equality does not necessarily hold in part a) of Theorem 3.7.2.

6. When does equality hold in part a) of Theorem 3.7.2? Is that "iff"?

7. Let $f : A \to B$, and $T \subset B$. Prove that $f_*(f^*(T)) = T \cap \text{Ran}(f)$.

8. Let $f : A \to B$, and suppose f is an injection. Prove that if $S \subset A$ then $f_*(A - S) = f_*(A) - f_*(S)$.

Reference

P. Halmos, *Naive Set Theory* (Van Nostrand Reinhold Co., New York, 1960)

Chapter 4

The Real Number System and Its Subsystems

Real numbers provide the means by which many quantities are measured. Whether it be the distance from point A to point B, the energy levels of a hydrogen atom, or the national budget, the real number system provides the idealized mathematical scale on which such measurements are made. In this chapter we investigate the main properties of the real number system and its subsystems. In the process we shall study the size of certain subsets of the set of real numbers, and we will use these as measures of the size of other sets.

4.1 Axioms for the Real Number System and Elementary Algebraic Properties

Our goal is to place the real number system, calculus, and the rest of mathematical analysis on a secure foundation. Even though it is not easy to say precisely what a real number is, it is easy to state precisely what our initial assumptions or axioms are. Once these axioms have been stated, we then demonstrate how the rest of the familiar properties of the real numbers can be deduced from these axioms. In this way mathematical analysis is given a firm foundation.

We know that we can add and multiply real numbers, so fundamentally the real number system is a set $\mathbb{R}$ together with two binary operations $+ : \mathbb{R} \times \mathbb{R} \to \mathbb{R}$, and $\times : \mathbb{R} \times \mathbb{R} \to \mathbb{R}$ which we call addition and multiplication. We write $x + y$ for $+(x,y)$, and xy for $\times(x,y)$. Our intuitive knowledge of the operations of addition and multiplication lead us to choose the following preliminary list of axioms.

The Field Axioms

A1) For all x and y in $\mathbb{R}$, $x + y = y + x$ (commutative law).
A2) For all x, y, and z in $\mathbb{R}$, $x + (y + z) = (x + y) + z$ (associative law).
A3) There exists a unique element called 0 in $\mathbb{R}$ such that for all x in $\mathbb{R}$, $x + 0 = 0 + x = x$. We call 0 the additive identity.
A4) For each x in $\mathbb{R}$ there exists a unique y in $\mathbb{R}$ such that $x + y = y + x = 0$. We denote this unique y by "$-x$". We call $-x$ the additive inverse of x.
A5) For all x and y in $\mathbb{R}$, $xy = yx$ (commutative law).
A6) For all x, y, and z in $\mathbb{R}$, $x(yz) = (xy)z$ (associative law).
A7) There is a unique element called 1 in $\mathbb{R}$ such that $1 \neq 0$, and for all x in $\mathbb{R}$, $1x = x1 = x$. We call 1 the multiplicative identity.
A8) For all x, y, and z in $\mathbb{R}$, $x(y + z) = xy + xz$ (distributive law).
A9) For each x in $\mathbb{R}$ with $x \neq 0$ there is a unique element y in $\mathbb{R}$ such that $xy = yx = 1$. We denote this element y by "x^{-1}". We call x^{-1} the multiplicative inverse of x.

Observe that the first theorems depend only on the axioms A1 through A8.

Theorem 4.1.1

For all x in $\mathbb{R}$, $x0 = 0$.

Proof:
Let $x \in \mathbb{R}$. We have $0 = 0 + 0$ (Additive identity). So, $x0 = x(0 + 0) = x0 + x0$ (Distribution). So,

$$0 = x0 + -(x0) = (x0 + x0) + -(x0) = x0 + (x0 + -(x0)),$$

(Additive inverse and associative law). Thus, $0 = x0 + 0 = x0$ (additive inverse and identity). •

Some remarks on this proof are in order. First, note the use of the associative law and the regrouping of parentheses in this proof. It can be proved from axioms A1 and A2 that numbers may be added in any order and in any grouping. For example: $(a + b) + (c + d) = ((a + b) + c) + d = d + ((c + a) + b)$. In view of this we shall often omit parentheses in such sums. The same goes for products. The proof of this "obvious" fact is quite difficult, and we will mention it again at the appropriate time.

Second, we have multiplied both sides of an equality, and we have added to both sides of an equality. Many beginners feel that an axiom is needed to justify these operations, but if the meaning of equality is recalled, it will be recognized that no such axiom is needed. Indeed, suppose $a = b$, then a and b denote the same number. Thus, $a + c$ is the sum of number a and number c, but then $b + c$ is the sum of the same two numbers. Alternatively, from the identity

$a + c = a + c$ we get $a + c = b + c$ by substitution. In addition, the above proof employs substitutions of equals for equals, this is also justified directly by the meaning of equality. Thus, if $a = b$ we may replace a by b in any formulas in which a occurs and vice versa.

Theorem 4.1.2

For all x in $\mathbb{R}$, $-(-x) = x$.

Proof:
Let $x \in \mathbb{R}$. We have $-x + x = 0$ (Why?) Thus, x is the unique additive inverse of $-x$. So, $x = -(-x)$. •

Theorem 4.1.3

For all x and y in $\mathbb{R}$, $(-x)y = -(xy) = x(-y)$.

Proof:
Let $x, y \in \mathbb{R}$. Now, $xy + (-x)y = [x + -x]y = 0y = 0$ (distributive law, additive inverse, and Theorem 4.1.1). Thus, $(-x)y = -(xy)$ (Why?). The other equality is proved similarly. •

Corollary 4.1.4

For all $x \in \mathbb{R}$, $(-1)x = -x$. Also, $(-1)(-1) = 1$.

Proof:
Let $x \in \mathbb{R}$. Then $(-1)x = -(1x) = -x$. Also, $(-1)(-1) = -(-1) = 1$. •

We remark that a corollary is a theorem that is obtained usually as an easy consequence of a preceding theorem or theorems.

The proof of the following theorem is left to the reader.

Theorem 4.1.5

For all x and y in $\mathbb{R}$, $(-x)(-y) = xy$.

Definition 4.1.6

We write $x - y$ for $x + -y$ and x/y or $\dfrac{x}{y}$ for xy^{-1}. Thus, $x - y = x + -y$, and $x/y = xy^{-1}$. •

Theorem 4.1.7

If $xy = 0$ then $x = 0$ or $y = 0$.

Proof:
Suppose $xy = 0$. If $x = 0$ we are done, so suppose $x \neq 0$. Then, $x^{-1}xy = x^{-1}0 = 0$ (What theorem?). But $x^{-1}xy = 1y = y$. So, $y = 0$. •

In the statement of Theorem 4.1.7 the universal quantifiers have been omitted. This is very common practice. The convention is that an omitted quantifier is to be taken to be universal.

The proof of the following theorem is left to the reader.

Theorem 4.1.8

1) $x(y - z) = xy - xz$.

2) If $x \neq 0$ and $z \neq 0$ then $\left(\frac{w}{x}\right)\left(\frac{y}{z}\right) = \frac{(wy)}{(xz)}$.

3) If $x \neq 0$ and $z \neq 0$ then $\frac{w}{x} + \frac{y}{z} = \frac{wz + xy}{xz}$.

The Order Axiom

Consider $\mathbf{Z}_3$, the equivalence classes of integers modulo 3 with the addition and multiplication of classes as defined in Chapter 3. It is easy to verify that this system satisfies axioms A1 through A9. For example the multiplicative inverse of [2] is [2], since [2][2] = [4] = [1], and [1] is clearly the multiplicative identity. The axioms A1 – A9 are a long way from characterizing the real number system. As we see, these axioms are properties that can be shared by systems different from the real number system. Systems that satisfy axioms A1 through A9 are called fields. It is our aim to add to these axioms until only one field remains that satisfies all the axioms – the real number field.

A10) There is a set $\mathbf{P} \subset \mathbb{R}$ such that:

a) for all x and y in $\mathbf{P}$, $x + y \in \mathbf{P}$, and $xy \in \mathbf{P}$, and

b) for any $x \in \mathbb{R}$, then exactly one of the following holds $x = 0$, or $x \in \mathbf{P}$, or $-x \in \mathbf{P}$. We call $\mathbf{P}$ the set of positive elements. We call b) the trichotomy law.

Note that 1 is in $\mathbf{P}$ because $1 \neq 0$ and if 1 is not in $\mathbf{P}$ then -1 is in $\mathbf{P}$, but $(-1)(-1) = 1$, so 1 would be in $\mathbf{P}$.

We shall say that a field that satisfies A10 is an ordered field. Note that $1 \neq 1 + 1 \neq 1 + 1 + 1 \neq$, etc, so $\mathbf{P}$ must contain infinitely many numbers. Thus finite fields like $\mathbf{Z}_3$ are not ordered fields. We have cut down on the number of systems that satisfy our axioms.

Definition 4.1.9

If $y - x \in \mathbf{P}$ we write $x < y$ and say "x is less than y". We may also write $y > x$ for $x < y$ and read "y is greater than x". We shall also write $x \leq y$ for $x < y$ or $x = y$. •

Theorem 4.1.10

If $x < y$ then $x + z < y + z$.

Proof:
Suppose $x < y$. Then $y - x \in \mathbf{P}$, so $(y + z) - (x + z) \in \mathbf{P}$, and so $x + z < y + z$. •

Theorem 4.1.11

If $x < y$ and $0 < z$ then $xz < yz$.

Proof:
Suppose $x < y$ and $0 < z$. Then, $y - x \in \mathbf{P}$, and $z \in \mathbf{P}$, and so $(y - x)z \in \mathbf{P}$. So, $yz - xz \in \mathbf{P}$, and so $xz < yz$. •

Theorem 4.1.12

If $x < y$ and $z < 0$ then $yz < xz$.

Proof:
Suppose $x < y$ and $z < 0$. Then $y - x \in \mathbf{P}$, and $-z \in \mathbf{P}$. So, $(y - x)(-z) \in \mathbf{P}$. But $(y - x)(-z) = y(-z) + (-x)(-z) = xz + (-yz) = xz - yz$. Thus, $yz < xz$. •

We close this section with the definition of absolute value.

Definition 4.1.10

Let $x \in \mathbb{R}$. We define $|x|$ to be x if $x \geq 0$ and $-x$ if $x < 0$. We call $|x|$ the absolute value of x. •

We leave the proof of some basic properties of the absolute value to the exercises.

Exercises for Section 4.1

In the following proofs only the above axioms, the convention on dropping parentheses, or previously proved results may be used.

1. Prove Theorem 4.1.5.

2. Prove that if $a + c = b + c$ then $a = b$.

3. Prove that if $ac = bc$ and $c \neq 0$ then $a = b$.

4. Prove Theorem 4.1.8.

5. Prove that if $a < b$ and $b < c$ then $a < c$.

6. Prove that if $a \leq b$ and $b \leq a$ then $a = b$.

7. Prove that $a = b$, or $a < b$, or $b < a$.

8. Prove that if $a^2 = aa$, then $a^2 \geq 0$.

9. Prove that if $x + z < y + z$ then $x < y$.

10. Prove that if $xz < yz$ and $z > 0$ then $x < y$.

11. Prove that if $x < y$ and $x > 0$ and $y > 0$ then $1/y < 1/x$.

12. Prove that,
    a) $|x| \geq 0$. b) $|xy| = |x||y|$. c) $|x + y| \leq |x| + |y|$. d) $||x| - |y|| \leq |x - y|$.

13. Show that $\mathbf{Z}_2$, $\mathbf{Z}_5$, and $\mathbf{Z}_7$ are fields, but that $\mathbf{Z}_4$, and $\mathbf{Z}_6$ are not. Make a conjecture.

4.2 The Completeness of the Reals

In this section we introduce the final axiom necessary for the characterization of the real numbers. We begin by discussing the need for this axiom.

We shall give a precise definition of the positive integers in the next section, but for the moment the following definition of the positive integers will do. The positive integers are: 1, 1 + 1, 1 + 1 + 1, etc. We denote these elements by 1, 2, 3, etc. The integers are 0, 1, −1, 2, −2, 3, −3, etc. The rational numbers are defined to be fractions m/n where m and n are integers, and $n \neq 0$. In section 4.3

we shall prove the intuitively clear proposition that the sums and products of integers are again integers. Accepting this, it is clear that sums and products of rational numbers are rational (use Theorem 4.1.8). Note that each rational number, except 0, has a multiplicative inverse. Indeed, if $m/n \neq 0$ then $m \neq 0$, so $(m/n)(n/m) = 1$. The positive rational numbers are those rationals equal to rationals of the form m/n where m and n are positive integers. Clearly, the rational numbers form an ordered field.

The axioms of an ordered field are not sufficient to form a foundation for the real numbers, for the real numbers have important properties that the rational numbers do not. One of the most important properties of the real numbers is that they cannot be divided into two nonempty sets, L and R, such that for all x in L and for all y in R, $x < y$ without L containing a maximum or R containing a minimum. This is clearly not the case for the rationals. Indeed, we shall show that $\sqrt{2}$ is irrational in section 4.6, so let

$$L = \{r \mid r \text{ is rational and } r < \sqrt{2}\}, \text{ and}$$
$$R = \{r \mid r \text{ is rational and } \sqrt{2} < r\}.$$

A pair of nonempty sets, L and R, such that for all x in L and for all y in R, $x < y$ and $L \cup R = \mathbb{R}$ is called a cut. Note that $L \cap R = \phi$.

We shall use the usual notation for intervals of reals, using "(" and ")" to indicate the omission of an end point and "[" and "]" for inclusion.

In terms of the usual picture of the real line, the only possible cuts are as shown in the following figures.

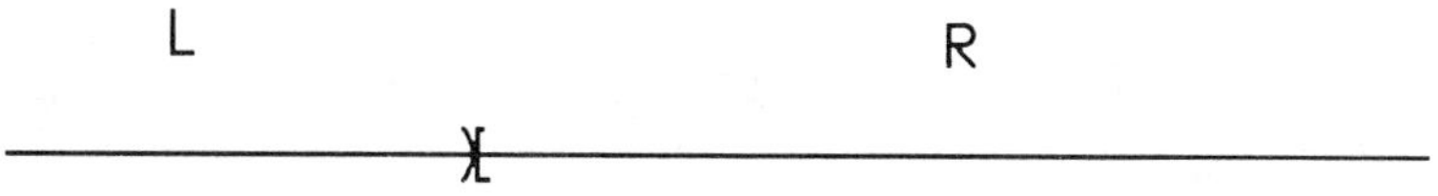

The Completeness Axiom

To state the axiom of completeness we need to introduce some terminology.

Definition 4.2.1

Let S be a subset of $\mathbb{R}$. Let $b \in \mathbb{R}$. We say b is an upper bound for S if and only if $x \leq b$ for all x in S. We say c is a least upper bound for S if and only if c is an upper bound for S, and for any other upper bound b of S, $c \leq b$. We often call a least upper bound of S a supremum or sup of S. •

We leave it to the reader to prove that the supremum of a set is unique. Given this we shall often write sup(S) or lub(S) for the supremum or least upper bound of S if it exists.

A11) Let S be a nonempty subset of $\mathbb{R}$. If S has an upper bound then
S has a least upper bound.

This axiom is called the axiom of the completeness of the real number system, and it is the axiom that distinguishes the reals from the rationals.

Theorem 4.2.2

Axiom A11 holds if and only if for every cut L and R of $\mathbb{R}$, $L = (-\infty,a]$ or $R = [a,\infty)$ for some $a \in \mathbb{R}$.

Proof:
Suppose A11 holds. Let L and R be a cut. Recall that L and R are not empty. Then L has an upper bound (Why?). Let a be the least upper bound of L. Then $a \in L$ or $a \in R$, since $L \cap R = \phi$ and $L \cup R = \mathbb{R}$.

Suppose $a \in L$. Now if $x \in L$ then $x \leq a$, so $x \in (-\infty,a]$. So $L \subset (-\infty,a]$. If $x \in (-\infty,a]$ and $x \in R$, then $x > a$ which is not so. Thus, $(-\infty,a] \subset L$. So, $L = (-\infty,a]$.

Suppose $a \in R$, then $a \leq y$ for all $y \in R$, for otherwise a could not be the sup of L. So, $R \subset [a,\infty)$. If $y \in [a,\infty)$ then $y \in R$, for otherwise $y < a$. So, $[a,\infty) \subset R$, and $R = [a,\infty)$.

Conversely, suppose for every cut L and R of $\mathbb{R}$, $L = (-\infty,a]$ or $R = [a,\infty)$ for some $a \in \mathbb{R}$. Let S be a nonempty set that has an upper bound, say b. Let $L = \{z \mid \text{there is an } x \text{ in } S \text{ such that } z \leq x\}$, then $L \neq \mathbb{R}$. Let $R = \mathbb{R} - L$. We now show that L and R form a cut. If $r \in L$ and $s \in R$ then we cannot have $s \leq r$, for then s would be in L. So, $r < s$, and we have a cut. Therefore, there exists an $a \in \mathbb{R}$ such that $L = (-\infty, a]$ or $R = [a, \infty)$. Now $S \subset L$, and so in either case a is an upper bound for S. Let c be an upper bound and suppose $c < a$. Then $(a + c)/2$ is an upper bound for S strictly bigger than c and so $(a + c)/2$ is in R and is strictly less than a (see Exercise 2). But that is impossible. So, $a \leq c$. Thus, a is the sup of S. •

The supremum form of the axiom of completeness, rather than the equivalent cut form, has been chosen because that is the form in which we shall most often use it.

The rest of this chapter is devoted to developing the consequences of the axioms A1 through A11. It is hoped that the reader will be impressed by the immense riches that flow by logical deduction from our few axioms. In Section 4.5 we note the way in which these 11 axioms characterize the real number system.

Exercises for Section 4.2

1. Prove that if a set $S \subset \mathbb{R}$ has a supremum then the supremum is unique.

2. In the proof of Theorem 4.2.2 we tacitly assumed that if $a < b$ then $a < (a + b)/2 < b$. Prove this.

3. Decide if S has a least upper bound. If so, what is it?
   a) $S = \mathbb{R}$. b) $S = (-3, 7)$. c) $S = (-\infty, 0]$. d) $S = (0, \infty)$.
   e) $S = \{x \mid x^2 < 4\}$. f) $S = \{x \mid x^2 < 2\}$.

4. Give a sensible definition of "lower bound" and "greatest lower bound". We often call the greatest lower bound of a set S the infimum and denote it by $\inf(S)$ or $\text{glb}(S)$.

5. Prove that if a nonempty set S has a lower bound it has a unique greatest lower bound.

6. What can you say about upper bounds of ϕ?

4.3 The Natural Numbers, Integers, and Induction

Everyone is familiar with the natural numbers 1, 2, 3, 4, etc. Our first task in this section is to define these familiar numbers within the framework of the axioms A1 through A11. In the last section we tentatively defined the natural numbers as 1, 1 + 1, 1 + 1 + 1, etc. This is not completely satisfactory, for in essence it involves counting which is an algorithm which is not described in our axioms. We need a definition of the natural numbers that lies within our axiom system. Still, the core of the idea is in our tentative definition. The idea is that the natural numbers are completely accounted for by adding 1 to itself repeatedly. However, we must capture the idea of repetition logically and without introducing counting. Here is how it is done.

Definition 4.3.1

Let $S \subset \mathbb{R}$. We say S is an inductive set if and only if

1) $1 \in S$, and
2) for all x in $\mathbb{R}$, if $x \in S$ then $x + 1 \in S$. •

Clearly, $\mathbb{R}$ is inductive as is **P** and many other subsets of $\mathbb{R}$. Intuitively it is clear that the natural numbers are inductive. Indeed, a little reflection shows that the set of natural numbers, as we intuitively understand it, contains no inductive subsets. This motivates the following definition.

Definition 4.3.2

The set $\mathbf{N} = \bigcap\{S \mid S \text{ is an inductive set}\}$ is called the set of natural numbers. •

Note that **N** is an inductive set. In fact **N** may be thought of as the smallest inductive set in $\mathbb{R}$.

We shall prove some familiar properties of the elements of this set. Most of the proofs will have a common form due to the way in which **N** is defined. This common form is called proof by induction which we now describe.

Proof by Induction

Suppose we wish to prove a proposition of the form:

"for every $n \in \mathbf{N}$, $P(n)$".

We first show that $P(1)$ is true, then we prove that for all $n \in \mathbf{N}$, if $P(n)$ is true then $P(n + 1)$ is true. We then conclude that for every $n \in \mathbf{N}$, $P(n)$ is true. Why is this valid?

Let $S = \{n \mid n \in \mathbf{N} \text{ and } P(n) \text{ is true}\}$. In showing that $P(1)$ is true we show that $1 \in S$. In showing that, for all $n \in \mathbf{N}$, if $P(n)$ is true then $P(n + 1)$ is true, we show that, for all n, if $n \in S$ then $n + 1 \in S$. Therefore we have shown that S is inductive. But, $S \subset \mathbf{N}$ by the definition of S. However,

$$\mathbf{N} = \bigcap\{S \mid S \text{ is an inductive set}\},$$

and so $S = \mathbf{N}$ by the definition of $\mathbf{N}$.

Consider two examples.

Example

For all $n \in \mathbf{N}$, $2 \mid n^2 + n + 2$.

Proof:
Let $P(n) = 2 \mid n^2 + n + 2$. Let $n = 1$. Then $n^2 + n + 2 = 4$. So $P(1)$ is true. Suppose $P(n)$ is true, then $2 \mid n^2 + n + 2$. Consider $n + 1$. We have, $(n + 1)^2 + (n + 1) + 2 = n^2 + 2n + 1 + n + 1 + 2 = n^2 + n + 2 + 2n + 2 = 2k + 2(n + 1) = 2(k + n + 1)$. So $P(n + 1)$ is true. Therefore, $P(n)$ is true for all n in $\mathbf{N}$. •

Example

For all $n \in \mathbf{N}$, $1 + 2 + \ldots + n = n(n + 1)/2$

Proof:
Let $n = 1$. Then $n(n + 1)/2 = 2/2 = 1$. So the proposition is true for 1. Now suppose it is true for n, and consider $n + 1$. Then,

$$\begin{aligned} 1 + 2 + \ldots + n + (n + 1) &= n(n + 1)/2 + (n + 1) \\ &= (n + 1)[n/2 + 1] \\ &= (n + 1)[n + 2]/2 \\ &= (n + 1)[(n + 1) + 1]/2. \end{aligned}$$

Therefore the proposition is true for $n + 1$, and hence for any n. •

The reader may wonder what the expression $1 + 2 + \ldots + n$ means. More generally, one can ask what $\sum_{i=1}^{n} a_i = a_1 + a_2 + \ldots + a_n$ means. We show how to use induction to define these and other expressions in due course. Meanwhile, we use induction to prove four elementary theorems about the natural numbers.

Theorem 4.3.3

For all $n \in \mathbf{N}$, $1 \leq n$.

Proof:
Certainly, $1 \leq 1$. Let $n \in \mathbf{N}$, and suppose $1 \leq n$. Then, $1 + 1 \leq n + 1$, and so $1 \leq n + 1$. (Why?) Therefore, for all n in N, $1 \leq n$. •

We say a natural number n is even if there is a natural number k such that $n = 2k$ where $2 = 1 + 1$. We say n is odd if there is a natural number k such that $n = 2k - 1$.

Theorem 4.3.4

For every n in N, n is even or odd, but not both.

Proof:
Let $n = 1$. Then $n = 2(1) - 1$, so 1 is odd. Now if 1 is also even, then $1 = 2k$ for some k in N. But $1 \leq k$, and $k < k + k = 2k$, so $1 < 2k$. So, 1 is not even.

Now suppose the theorem is true for n. Consider $n + 1$. If n is even then $n = 2k$ and so $n + 1 = 2k + 1 = 2(k + 1) - 1$. Thus $n + 1$ is odd. If n is odd then $n = 2k - 1$, and so $n + 1 = 2k - 1 + 1 = 2k$. So, $n + 1$ is even. Now if $n + 1$ is even and odd then $n + 1 = 2k_1$, and $n + 1 = 2k_2 - 1$, and so n is both even and odd. That contradicts the assumption on n, so $n + 1$ is not both even and odd. •

Many readers may feel that the above proofs are belabouring the obvious. Readers who feel this way should examine their souls with the following question. "Are these theorems obvious, or am I rejecting the proofs on the basis that I have known these theorems as facts for a long time?" If this question hits close to home, remember that we are trying to establish the properties of the natural numbers on the basis of the axioms and definitions. Deduction from the axioms is the name of the game. We are not engaged in a mere review of familiar facts. The reader may take some comfort in the fact that, in the early development of properties in an axiom system, proving simple things can be frustratingly tricky.

Theorem 4.3.5

For all m and n in N, $m + n \in \mathbf{N}$, and $mn \in \mathbf{N}$.

Proof:
Let $m \in \mathbf{N}$. Now if $n = 1$ then $m + n = m + 1$. So, $m + n \in \mathbf{N}$. Now suppose that $m + n \in \mathbf{N}$, then $m + (n + 1) = (m + n) + 1$, so $m + (n + 1) \in \mathbf{N}$. Therefore, by induction it follows that for all n in N, $m + n \in \mathbf{N}$. But m was arbitrary, so universally generalizing gives the first part of the theorem, for all m in N and for all n in N, $m + n \in \mathbf{N}$. We leave the second part to the reader. •

This proof contains a method that is worth further comment. We needed to prove a statement of the form "For every m in **N**, and for every n in **N**, $P(m,n)$". The strategy was to universally specify m, and then prove that "for every n in **N**, $P(m,n)$" is true by induction. Noting that the induction did not depend on m, we then universally generalized on m to get the result.

The proof of the next theorem is surprisingly subtle.

Theorem 4.3.6

For all m and n in **N**, if $m < n$ then $n - m$ is in **N**.

Proof:
We induct on m. Suppose $m = 1$. Then we must prove that for all n in **N**, if $1 < n$ then $n - 1$ is in **N**. We do this by induction on n. If $n = 1$ then we are done (Why?). Suppose it is true for n. Consider $n + 1$. Certainly $1 < n + 1$. But then $n = n + 1 - 1$, and so $n + 1 - 1$ is in **N**. So it is true for all n. Thus, the theorem is true for $m = 1$.

Suppose the theorem is true for m. That is, suppose that

for all n in **N**, if $m < n$ then $n - m$ is in **N**.

Consider $m + 1$. If $m + 1 < n$, then $m + 1 < n + 1$, and so $m < n$. Thus, $n - m$ is in **N**. Now $n - m > 1$, and so $n - m - 1 = n - (m + 1)$ is in **N**, since the theorem has been proved for the case $m = 1$. Thus, the theorem follows by induction. •

Definition by Induction

Many expressions in mathematics such as $\sum_{i=1}^{n} a_i$ and a^n are defined by the use of an algorithm. For example we often see $\sum_{i=1}^{n} a_i$ defined to be the sum $a_1 + a_2 \ldots + a_n$. The algorithm is to add the numbers $a_1, a_2, \ldots$, until we reach the number a_n and then stop. However, such an algorithm is not explicitly sanctioned by our axioms for the real number system since the notion of counting is not explicit in the axioms. We now show how induction can be used to remedy this deficiency.

Definition 4.3.7

Let a_i be a real number for each $i \in \mathbf{N}$. Define $\sum_{i=1}^{1} a_i = a_1$, and define

$$\sum_{i=1}^{n+1} a_i = \sum_{i=1}^{n} a_i + a_{n+1}. \bullet$$

Note that if we ask if this definition has assigned a real number to the symbol $\sum_{i=1}^{n} a_i$ for each n, a proof by induction immediately gives the answer – yes! Also note that this definition does not employ the concept of an algorithm in the definition, but the algorithm is implied by the definition. For example, $\sum_{i=1}^{1} a_i =$ a_1, $\sum_{i=1}^{2} a_i - \sum_{i=1}^{1} a_i + a_2 = a_1 + a_2$, and $\sum_{i=1}^{3} a_i = \sum_{i=1}^{2} a_i + a_3 = (a_1 + a_2) + a_3$.

Finally, we remark that it can be shown that this definition of $\sum_{i=1}^{n} a_i$ can be used to show that the sum of a collection of numbers can be grouped with parentheses in any manner without affecting the sum. The proof is a little too complicated to include here, but we refer the interested reader to the reference by Zariski and Samuel listed at the end of this chapter. Familiar properties of these sums are established in the exercises.

Consider another example of definition by induction. If n is a natural number what does a^n stand for? The answer from elementary algebra is,

$$a^n = aaa \ldots a \ \{n \text{ times}\}.$$

Again, this introduces an algorithm into the definition that is unnecessary. Here is how definition by induction handles it.

Definition 4.3.8

Let $a \in \mathbb{R}$. Define $a^1 = a$, and for all $n \in \mathbf{N}$, define $a^{n+1} = a^n a$. •

We can now prove many familiar facts about powers. Here is one of them.

Example

For all m and n in N, $a^{m+n} = a^m a^n$.

Proof:
Let $m \in$ N. If $n = 1$ then $a^{m+1} = a^m a = a^m a^1$. Now suppose that $a^{m+n} = a^m a^n$. Then $a^{m+(n+1)} = a^{(m+n)+1} = a^{m+n}a = a^m a^n a = a^m a^{n+1}$. So, for any fixed m the result holds for all n. Generalizing, it holds for all m and n. •

In a similar manner, general products $\prod_{i=1}^{n} a_i$ which are defined algorithmically to be the product $a_1 a_2 \ldots a_n$ can be defined by induction.

Well-Ordering and Strong Induction

We close this section with two theorems which are basic tools in many proofs that involve statements about the natural numbers.

Definition 4.3.9

If $S \subset$ N, and there is an $n_0 \in S$ such that $n_0 \leq k$ for all $k \in S$, we call n_0 the least element of S. •

Theorem 4.3.10 (The Well-Ordering Principle)

If $S \subset$ N, and $S \neq \phi$, then S contains a least element.

Proof:
Suppose S has no least element. Let $T =$ N $- S$. Then $1 \in T$ (Why?). Suppose, for any n, $\{1,2, \ldots, n\} \subset T$, and consider $n + 1$. Then $n + 1$ must be in T, for otherwise $n + 1$ would be the least element of S. So,

$$\{1,2, \ldots, n + 1\} \subset T.$$

Thus, for all n, $\{1, 2, \ldots, n\} \subset T$, so $T =$ N. But then $S = \phi$. That is a contradiction. •

In the proof above the truth for $n + 1$ depends not only on the truth for n but for all of the predecessors of n. The need to "look back" further than n in a proof by induction is quite common, and so we state and prove a version of proof by induction that deals with this situation.

Theorem 4.3.11 (Strong Induction)

Suppose that $k_0 \in \mathbf{N}$, and suppose that
1) $P(k_0)$ is true, and
2) for all $n \in \mathbf{N}$, $n \geq k_0$, $P(k)$ is true for all
$k \in \{k_0, k_0 + 1, k_0 + 2, \ldots, n\}$ implies that $P(n + 1)$ is true.
Then, $P(n)$ is true for all $n \geq k_0$.

Proof:
Assume hypotheses 1) and 2). Suppose $P(n)$ is not true for some $n \geq k_0$. Let $T = \{n \mid P(n) \text{ is false}\}$. Then $T \neq \phi$. So T has a least element, say n_0, with $n_0 > k_0$. Then, $P(k)$ is true for all

$$k \in \{k_0, k_0 + 1, k_0 + 2, \ldots, n_0 - 1\}.$$

But by hypothesis 2) $P(n_0)$ is true, and this is a contradiction. •

Example

Let $a_1 = 1$, and $a_2 = 2$. Let $a_n = a_{n-1} + a_{n-2}$ for $n \geq 3$. (definition by induction!) Then $a_n \leq (7/4)^n$ for all $n \geq 3$.

Proof:
For $n = 3$, $a_n = a_1 + a_2 = 3 \leq (7/4)^3$. Now suppose the statement is true for all $k \in \{3, 4, \ldots, n\}$. Then $a_{n+1} = a_n + a_{n-1}$, and so $a_{n+1} \leq (7/4)^n + (7/4)^{n-1}$. But, $(7/4)^n + (7/4)^{n-1} = (7/4)^{n-1}[\frac{7}{4} + 1] = (7/4)^{n-1}\frac{11}{4}$.
But, $\frac{11}{4} \leq (7/4)^2$. So. $a_{n+1} \leq (7/4)^{n+1}$. •

We conclude this section with the definition of the integers. Define,

$$\mathbf{Z} = \{x \mid x \in \mathbf{N}, x = 0, \text{ or } -x \in \mathbf{N}\}.$$

It is trivial to prove that sums and products of integers are integers. Indeed, we have proved that the sum and product of natural numbers are natural numbers. The extension to the integers is then made by looking at the possible cases.

Exercises for Section 4.3

Assume any needed properties of the integers.
Use induction, not strong induction, to prove the following.

1. For all n in $\mathbf{N}$, $2^n \geq n$.
2. For all n in $\mathbf{N}$, $1^2 + 2^2 + \ldots + n^2 = \dfrac{n(n+1)(2n+1)}{6}$.
3. Let $m \in \mathbf{N}$, and $G_m = \{k \mid k \geq m, \text{ and } k \in \mathbf{N}\}$. Prove that if P($m$) is true, and for every $k \in G_m$, if $P(k)$ is true then $P(k+1)$ is true, then $P(k)$ is true for all $k \in G_m$. Use this to prove:
   a) For all $n \in G_3$, $9 \mid [n^3 + (n-1)^3 + (n-2)^3]$.
   b) For all $n \in G_2$, $133 \mid (11^n + 12^{2n-3})$.

4. Use definition by induction to define $n! = 1 \times 2 \times \ldots \times n$.

5. Complete the proof of Theorem 4.3.5.

6. Use Definition 4.3.7 to prove that $\sum_{i=1}^{n}(a_i + b_i) = \sum_{i=1}^{n} a_i + \sum_{i=1}^{n} b_i$, and

   $\sum_{i=1}^{n} ca_i = c\sum_{i=1}^{n} a_i$ for all $n \in \mathbf{N}$.

7. Let $a \in \mathbb{R}$. For all m and n in $\mathbf{N}$, prove that $a^{mn} = (a^m)^n$.

8. Let $a \in \mathbb{R}$, and $a \neq 0$. Define $a^0 = 1$, and for $n \in \mathbf{N}$ define $a^{-n} = 1/a^n$. Prove that for all m and n in $\mathbf{Z}$ that $a^{m+n} = a^m a^n$, and $a^{mn} = (a^m)^n$.

9. Use Strong Induction to prove that if $a_1 = 1$, $a_2 = 2$, $a_3 = 3$, and $a_n = a_{n-1} + a_{n-2} + a_{n-3}$ for $n > 3$, then $a_n \leq 3^n$ for all $n \geq 3$.

10. Suppose you only have three dollar and five dollar bills. What is the largest debt you cannot pay with these bills. Use strong induction to prove your conjecture.

11. Prove the Binomial Theorem. Namely, for all $n \in \mathbf{N}$,

    $(a+b)^n = \sum_{i=0}^{n} C(n,k)a^{n-k}b^k$, where $C(n,k) = n!/[k!(n-k)!]$. Note

    that $\sum_{i=0}^{n} a_i = a_0 + \sum_{i=1}^{n} a_i$. .

12. Let $k_0 \in \mathbf{Z}$. Prove that $S = \{k \mid k \in \mathbf{Z} \text{ and } k \geq k_0\}$ is well-ordered.

4.4 Divisors, Primes, GCDs, and the Division Algorithm

In this section we derive several important properties of the integers.

We begin this section with the divisibility properties of integers. We shall allow ourselves to use any basic algebraic property of the integers that we need even though a formal proof may not have been given. Readers should supply such proofs when needed.

We begin with the definition of "divides".

Definition 4.4.1

Let a and b be integers. We say a divides b if and only if there is an integer c such that $b = ac$. In this case we write $a \mid b$, and we call a a divisor of b.•

Note that $a \mid b$ is a statement, not a fraction.

Examples

1) $3 \mid 6$ because $6 = 3 \times 2$.
2) $-4 \mid 12$ because $12 = (-3)(-4)$. •

We now prove a few simple theorems on divisibility.

Theorem 4.4.2

Suppose a, b, and c are in **Z**. If $a \mid b$ and $b \mid c$ then $a \mid c$.

Proof:
Suppose $a \mid b$ and $b \mid c$. Then there are integers k_1 and k_2 such that $b = ak_1$ and $c = bk_2$. So, $c = ak_1k_2$. So, $a \mid c$. •

Theorem 4.4.3

Suppose $a \mid b$ and $a \mid c$ then $a \mid (b + c)$ and $a \mid (b - c)$.

Proof:
We have $b = ak_1$ and $c = ak_2$, so $b + c = a(k_1 + k_2)$, so $a \mid (b + c)$. The proof that $a \mid (b - c)$ is similar. •

In the next theorem we make use of the absolute value, so we recall that $|a| = a$ if $a \geq 0$, and $|a| = -a$ if $a < 0$.

Theorem 4.4.4

If $a \mid b$, and $b \neq 0$, then $|a| \leq |b|$.

Proof:
Let $b = ak$, so $|b| = |a||k|$. Now $|b| \neq 0$, so $|a| \neq 0$, and $|k| \neq 0$. So $1 \leq |k|$, and so $|a| \leq |a||k| = |b|$. •

Whether a divides b or not, we can always perform long division to get a quotient and remainder. That this can be done is our next theorem.

Theorem 4.4.5 (The Division Algorithm)

Let $a \in \mathbf{Z}$, and $b \in \mathbf{N}$. Then there exist unique integers q and r such that

$$a = bq + r, \text{ and } 0 \leq r < b.$$

Of course, q is called the quotient and r is called the remainder.

Proof:
Let $S = \{y \mid y \in \mathbf{Z}, y \geq 0, \text{ and } y = a - bx \text{ for some } x \in \mathbf{Z}\}$. Since $b \geq 1$, $(-|a|b) \leq -|a| \leq a$. Thus, $a - (-|a|b) \geq 0$, so setting $x = -|a|$ and $y = a - bx$ we see that S is not empty. By the Well-Ordering Principle S has a least element, say r. Thus, $r = a - bq$ for some q, and $r \geq 0$ by the definition of S. So $a = bq + r$, and $0 \leq r$.

We show that $r < b$. Suppose not, then $b \leq r$. Therefore,

$$a - bq - b = r - b, \text{ and so}$$
$$a - b(q + 1) = r - b \geq 0.$$

Hence, $r - b \in S$. But this is impossible since r is the least member of S. So, $r < b$.

Finally, we prove that the q and r are unique. Suppose,

$$a = bq_1 + r_1 = bq_2 + r_2$$

with $0 \leq r_1 \leq r_2 < b$. Then $b|q_1 - q_2| = r_2 - r_1$. But $b \geq 1$, so if $q_1 - q_2 \neq 0$ then $r_2 - r_1 \geq b$. That is impossible. So, $q_1 = q_2$, and $r_1 = r_2$. •

Base Ten Representations

As an application of the division algorithm we shall show how every natural number can be represented in the familiar base ten notation. In this way we shall tie the reader's past experience to the current discussion.

Define the symbols 2, 3, . . . , 10 to be (1 + 1), (1 + 1 + 1), etc.

Theorem 4.4.6

For every $n \in \mathbf{N}$ there is a unique nonnegative integer k and unique nonnegative integers $a_0, a_1, \ldots, a_k$ where $a_k \neq 0$, $a_i = 0, 1, 2, \ldots,$ or 9, and

$$n = \sum_{i=0}^{k} a_i 10^i = a_k 10^k + a_{k-1} 10^{k-1} + \ldots + a_1 10 + a_0.$$

We call $a_k 10^k + a_{k-1} 10^{k-1} + \ldots + a_1 10 + a_0$ the representation of n in base ten.

Proof:
We induct on n. For $n = 1$, $k = 0$, and $a_0 = 1$. Clearly, this is the unique representation of 1. Suppose the theorem is true for any q such that $0 \leq q < n$. We divide n by 10 to get $n = 10q + r$ where $0 \leq r < 10$ and $0 \leq q < n$. But by the induction hypothesis

$$q = a_k 10^k + \ldots + a_1 10 + a_0,$$

and so,

$$n = 10q + r = a_k 10^{k+1} + \ldots + a_1 10^2 + a_0 10 + r.$$

Therefore, n has a representation in base ten.

Now suppose $n = a_k 10^k + \ldots + a_1 10 + a_0 = b_k 10^k + \ldots + b_1 10 + b_0$, and suppose $a_j \neq b_j$ for some j. Let i be the least integer such that $a_{k-i} \neq b_{k-i}$. Note that if the base ten representations do not have the same length we lengthen the shorter one by adding powers of 10 with zero coefficients. Cancelling gives

$$s = a_{k-i} 10^{k-i} + \ldots + a_1 10 + a_0 = b_{k-i} 10^{k-i} + \ldots + b_1 10 + b_0.$$

But $s < n$, so by induction these representations are the same so $a_{k-i} = b_{k-i}$, this is impossible, and so the representations of n are the same. •

We conclude our remarks about the representation of natural numbers in base ten by discussing the algorithms for the addition and multiplication of numbers in base ten.

The algorithm for addition is this. The base ten sum of two natural numbers is as follows:

$$(a_k 10^k + \ldots + a_1 10 + a_0) \;+\; (b_k 10^k + \ldots + b_1 10 + b_0)$$

$$= q_k 10^{k+1} + c_k 10^k + \ldots + c_1 10 + c_0,$$

where, from the division algorithm, $a_0 + b_0 = 10q_0 + c_0$, and $a_i + b_i + q_{i-1} = 10q_i + c_i$.

The algorithm just described is the familiar procedure for adding two numerals. The algorithm produces the unique base ten representation of the sum as we now prove by induction on the index k.

If $k = 1$ the assertion is immediate from the division algorithm. Suppose the algorithm produces the correct sum for natural numbers where the longer number has leading nonzero digit a_k. Now consider $k + 1$. We have

$$a_{k+1}10^{k+1} + a_k 10^k + \ldots + a_0 + b_{k+1}10^{k+1} + b_k 10^k + \ldots + b_0$$

$$= a_{k+1}10^{k+1} + b_{k+1}10^{k+1} + (a_k 10^k + \ldots + a_0) + (b_k 10^k + \ldots + b_0)$$

$$= a_{k+1}10^{k+1} + b_{k+1}10^{k+1} + q_k 10^{k+1} + c_k 10^k + \ldots + c_0$$

$$= (a_{k+1} + b_{k+1} + q_k)10^{k+1} + c_k 10^k + \ldots + c_0$$

$$= q_{k+1}10^{k+2} + c_{k+1}10^{k+1} + c_k 10^k + \ldots + c_0.$$

As for the product, we have

$$(a_k 10^k + \ldots + a_1 10 + a_0)(b_k 10^k + \ldots + b_1 10 + b_0)$$

$$= (a_k b_0 10^k + \ldots + a_1 b_0 10 + a_0 b_0) + (a_k b_1 10^{k+1} + \ldots + a_0 b_1) + \ldots + (a_k b_k 10^{2k} + a_{k-1} b_k 10^{2k-1} + \ldots + a_0 b_k).$$

We write each $(a_k b_i 10^{k+i} + \ldots + a_1 b_i 10^{1+i} + a_0 b_i 10^i)$ in its unique base ten represetation by again dividing by 10 and carrying. We add these base ten representations as we have described above to get the result. If readers multiply two numerals by the algorithm they learned in the elementary grades they will see that the above is a formal description of this procedure. We leave the reader to prove by induction that the algorithm produces the correct product.

In view of these remarks, we may think of the natural numbers as numerals in base ten as we so often do.

Primes and the Greatest Common Divisor

Of fundamental importance in number theory are the prime numbers which we now define.

Definition 4.4.7

A number $p \in \mathbf{Z}$ is a prime if and only if $p \neq \pm 1$, and p is divisible by ± 1 and $\pm p$ only. •

Examples

1) ± 2, ± 3, ± 5 ,± 7, etc are primes.
2) 6 is not prime since $2 \mid 6$. •

Theorem 4.4.8

Let m be an integer such that $m \neq \pm 1$. Then m has a prime divisor.

Proof:
If $m = 0$ then m has a prime divisor, say 2. If $m \neq 0$ let S be the set of positive divisors of m greater than 1. This is not empty since $|m| \mid m$. Let p be the least element in S. If p is not prime then $p = ab$ where $1 < a < p$. But then $a \mid m$, and so $a \in S$. This is not possible, since p was the least positive divisor of m. So p is a prime. •

How many prime numbers are there? Our next theorem answers this question.

Theorem 4.4.9

The set of primes is infinite.

Proof:
Suppose the set of primes is finite. We can list them as $p_1, p_2, \ldots, p_n$. Let $m = p_1 p_2 \ldots p_n + 1$. Then m has a prime divisor, say q. Now q is one of the

p_i, so $q \mid p_1 p_2 \ldots p_n$ and $q \mid m$. So $q \mid 1$. This is a contradiction, so the set of primes must be infinite. •

Given two numbers we can talk about their common divisors. Most important is the greatest common divisor.

Definition 4.4.10

Let a and b be nonzero integers. If $c \mid a$ and $c \mid b$, then we call c a common divisor of a and b. If d is a common divisor, and for every common divisor c, $c \leq d$, then we call d the greatest common divisor of a and b. We use the abbreviation gcd, and we write $d = (a,b)$. •

We have assumed that the gcd is unique, but this is obvious.

Theorem 4.4.11

Let a and b be nonzero integers. Then $d = (a,b)$ exists, and there are integers m_0 and n_0 such that $d = m_0 a + n_0 b$.

Proof:
Let $S = \{y \mid y \in \mathbf{Z}, y > 0, \text{ and } y = ma + nb \text{ for some } m, n \in \mathbf{Z}\}$. Clearly, S is not empty since $a^2 + b^2 \in S$. Let d be the least element of S. Then $d = m_0 a + n_0 b$ for some integers m_0 and n_0.

We now show that d is a common divisor of a and b. Suppose d does not divide a, then $a = dq + r$ where $0 < r < d$. Then $a = (m_0 a + n_0 b)q + r$, and so $r = (1 - m_0 q)a + (-n_0 q)b$. Therefore, $r \in S$. But $r < d$, so this is a contradiction, thus $d \mid a$. Similarly $d \mid b$. Finally, if $c \mid a$ and $c \mid b$ then $c \mid m_0 a + n_0 b$, and so $c \mid d$. But then $c \leq d$. •

Definition 4.4.12

Let a and b be nonzero integers. If $(a,b) = 1$ we say a and b are relatively prime. •

Theorem 4.4.13

Let a, b and c be nonzero integers, and $(a,c) = 1$. If $c \mid ab$ then $c \mid b$.

Proof:
There are integers m_0 and n_0 such that $1 = m_0 a + n_0 c$, so $b = m_0 ab + n_0 cb$. But c divides the two terms on the right, so $c \mid b$. •

We conclude the section with The Fundamental Theorem of Arithmetic.

Theorem 4.4.14

Let n be an integer with $n \geq 2$. Then there exists an $r \in \mathbf{N}$ and positive primes $p_1, p_2, \ldots, p_r$ such that $n = p_1 p_2 \ldots p_r$. Moreover, except for the order of the primes in the product, the factorisation is unique.

Proof:
We first prove the existence of a factorization, and then we prove the uniqueness. If $n = 2$ then n is factored as a product of primes. Suppose every integer less than n and greater than or equal to 2 has a prime factorization. If n is prime we are done. If n is not prime then $n = ab$ with $2 \leq a < n$, and $2 \leq b < n$. So a and b have prime factorizations, and so does n. So by strong induction every $n \geq 2$ has a prime factorization.

Now suppose $n = p_1 p_2 \ldots p_r$, and $n = q_1 q_2 \ldots q_s$ where the p_i and q_i are primes. Also assume that $p_i \leq p_{i+1}$ and $q_i \leq q_{i+1}$. If $p_1 \neq q_1$ then $(p_1, q_1) = 1$, and so $p_1 \mid q_2 \ldots q_s$, and $q_1 \mid p_2 \ldots p_r$. So, $p_1 \leq q_1$, and $q_1 \leq p_1$. So, $p_1 = q_1$.(Why?) Then we have $p_2 \ldots p_r = q_2 \ldots q_s$. Reasoning as before $p_2 = q_2$. Thus, r must equal s, and $p_i = q_i$ for each i. •

Exercises for Section 4.4

1. Long divide −58 by 7 finding the quotient and remainder.

2. Prove that if p is a prime and $p \mid ab$ then $p \mid a$ or $p \mid b$.

3. Prove that if b is a positive integer, $b \neq 1$, and b is not prime, then there exists a positive prime p such that $p \mid b$ and $p \leq b^{1/2}$.

4. Prove that if $d = (a,b)$ and $a = k_1 d$ and $b = k_2 d$, then $(k_1, k_2) = 1$.

5. Prove that if $(a,m) = (b,m) = 1$, then $(ab,m) = 1$.

6. Prove that if $a \mid m$ and $b \mid n$ then $(a,b) \mid (m,n)$.

7. Prove that the gcd (a,b) is calculated in the following way.
   For each prime that occurs in a or b, take that prime to the least power to which it occurs in either a or b. Take these primes to these powers and multiply them together. The result is (a,b). Note, some of the powers may be zero.

8. Prove that if $am_0 + bn_0 = 1$ then $(a,b) = 1$.

9. If $x = y + z$ then $(x,y) = (y,z)$ for any nonzero integers x, y, and z.

10. Prove by induction that the algorithm for the multiplication of natural numbers in base ten gives the correct product.

11. Repeat the base ten development for base 2.

4.5 Finite and Infinite Sets

In this section and Section 4.7 we consider some of the intriguing properties of infinite sets. We begin by considering the definition and properties of finite sets.

Definition 4.5.1

Let $\mathbf{N}_n = \{k \mid k \in \mathbf{N}, \text{ and } k \leq n\}$. A set S is finite if and only if $S = \phi$, or there is an $n \in \mathbf{N}$ and a bijection $f : \mathbf{N}_n \to S$. We say S is infinite otherwise. We call $\mathbf{N}_n$ an initial segment of the positive integers. •

This definition says that a set S is finite if its elements can be placed into one to one correspondence with an initial segment of the positive integers. For example, if S is the set of letters of the alphabet then S is finite. Indeed, $1 \to a$, $2 \to b, \ldots, 26 \to z$.

Our first theorem is a special case of the assertion that no finite set can be placed into one to one correspondence with a proper subset of itself.

Theorem 4.5.2

For any $n \in \mathbf{N}$, there is no injection $f : \mathbf{N}_n \to \mathbf{N}_n$ such that $\text{Ran}(f) \neq \mathbf{N}_n$.

Proof:
We procede by induction. Let $n = 1$. Then $\mathbf{N}_1 = \{1\}$. The only proper subset is ϕ, so there is no injection whose range is a proper subset of $\mathbf{N}_1$.

Now suppose the theorem is true for n. Consider $\mathbf{N}_{n+1}$. Suppose $S \subset \mathbf{N}_{n+1}$, and $S \neq \mathbf{N}_{n+1}$. Suppose $f : \mathbf{N}_{n+1} \to S$ is a bijection. We will derive a contradiction. There are two possibilities, $n + 1 \in S$, or $n + 1 \notin S$. If $n + 1 \notin S$, then $S \subset \mathbf{N}_n$. Define $g : \mathbf{N}_n \to S - \{f(n + 1)\}$ by $g(k) = f(k)$ for each $k \in \mathbf{N}_n$. Now g is one to one and onto $S - \{f(n + 1)\}$, and $S - \{f(n + 1)\}$

is a proper subset of $\mathbf{N}_n$. This is impossible by the induction hypothesis for n. So, $n + 1 \in S$.

We define a map $h : \mathbf{N}_n \to S - \{n + 1\}$ as follows. If $f(n + 1) = n + 1$, set $h(k) = f(k)$ as before. Then h is an injection, and $S - \{n + 1\}$ is a proper subset of $\mathbf{N}_n$, and we have a contradiction as before. If $f(n + 1) \neq n + 1$, then set $f(n + 1) = r_0$ and $f(s_0) = n + 1$ where $s_0 \in \mathbf{N}_n$. Now define $h(k) = f(k)$ if $k \neq s_0$ and $k \in \mathbf{N}_n$, and $h(s_0) = r_0$. Readers should draw a function diagram of the construction of h in order to see what the idea behind the argument is. We need to show that h is an injection. Suppose $h(k_1) = h(k_2)$. If $k_1 \neq s_0$ and $k_2 \neq s_0$ then $f(k_1) = f(k_2)$, so $k_1 = k_2$. If $k_1 = s_0$ and $k_2 \neq s_0$, then $h(k_1) = h(s_0) = r_0$. But $k_2 \neq s_0$, so $h(k_2) = f(k_2)$, and $k_2 \neq n + 1$. Thus, $h(k_2) \neq r_0$. That is a contradiction, so $k_1 = k_2$ again. The last case is the same. Therefore h is an injection. Now $S - \{n + 1\}$ is a proper subset of $\mathbf{N}_n$, and so we have our final contradiction. Thus, the theorem follows by induction. •

Corollary 4.5.3

If S is a finite set, and $S \neq \phi$, then there exists a unique n such that there is a bijection $f : \mathbf{N}_n \to S$.

Proof:
Suppose not. Suppose $n \neq m$, and $f : \mathbf{N}_n \to S$ and $g : \mathbf{N}_m \to S$ are bijections. Assume that $n < m$. Then $f^{-1} \circ g : \mathbf{N}_m \to \mathbf{N}_n$ is a bijection, $\mathbf{N}_n \subset \mathbf{N}_m$, and $\mathbf{N}_n \neq \mathbf{N}_m$. This is impossible by Theorem 4.4.2. •

In view of this we make the following definition.

Definition 4.5.4

If S is a finite set, and $S \neq \phi$, we call the unique n associated with S in the corollary the number of elements in S. •

It is intuitively plausible that every subset of a finite set is finite. We now prove this.

Theorem 4.5.5

Any subset of $\mathbf{N}_n$ is finite.

Proof:
We procede by induction on n. If $n = 1$ and $S \subset \mathbf{N}_1$, then $S = \mathbf{N}_1$ or $S = \phi$. In either case S is finite.

Suppose the theorem is true for n. Let $S \subset \mathbf{N}_{n+1}$. If $n + 1 \notin S$ then $S \subset \mathbf{N}_n$, and so S is finite. If $n + 1 \in S$, then $S - \{n + 1\} \subset \mathbf{N}_n$, and so $S - \{n + 1\}$ is finite. If $S - \{n + 1\} = \phi$, then $S = \{n + 1\}$, and that is obviously finite (Why?). If $S - \{n + 1\} \neq \phi$, then there is a bijection $f : \mathbf{N}_m \to S - \{n + 1\}$ with $m \leq n$. (Why?) Define $g : \mathbf{N}_{m+1} \to S$ by $g(k) = f(k)$ for $1 \leq k \leq m$, and $g(m + 1) = n + 1$. Then g is a bijection and S is finite. •

We leave the proof of the following corollaries to the reader.

Corollary 4.5.6

Any subset of a finite set is finite.

Corollary 4.5.7

If S is a finite set then there is no injection of S onto a proper subset of S.

Corollary 4.4.7 implies that if S is a set that has an injection whose range is a proper subset then S is infinite.

Definition 4.5.8

We say that a set S is Dedekind infinite if and only if there is an injection of S onto a proper subset of S. •

We have proved that Dedekind infinite implies infinite. We conclude this section with the converse.

Theorem 4.5.9

If a set S is infinite (i.e. not finite) then it is Dedekind infinite.

Proof:
Suppose S is infinite. We define an injection $f : \mathbf{N} \to S$ by induction. Now $S \neq \phi$, so choose $x_1 \in S$. Let $f(1) = x_1$. Suppose f has been definied on

$$\{1, 2, \ldots, n\}.$$

Since S is infinite $S - \{f(1), f(2), \ldots, f(n)\}$ is not empty. Choose x_{n+1} in $S - \{f(1), f(2), \ldots, f(n)\}$, and define $f(n + 1) = x_{n+1}$. Clearly, f is an injection (Why?). Now define $g : S \to S$ as follows. Let $g(x) = x$ for $x \in S - \text{Ran}(f)$, and for $x_i \in \text{Ran}(f)$ set $g(x_i) = x_{2i}$. Clearly, g is an injection and its range is not S.•

Exercises for Section 4.5

1. Prove Corollary 4.5.6.

2. Prove Corollary 4.5.7.

3. Let A be a finite set and B any set. Prove that $A \cap B$ is finite.

4. Let A and B be finite sets, and let $A \cap B = \phi$. Prove that $A \cup B$ is finite.

5. Let A and B be finite sets. Prove that $A \cup B$ is finite.

6. Prove that for any n, if $A_1, A_2, \ldots,$ and A_n are finite then $A_1 \cup A_2 \cup \ldots \cup A_n$ is finite.

7. If A and B are finite prove that $A \times B$ is finite.

4.6 The Rationals and the Reals

So far we have not employed the completeness axiom A11. In this section we use the completeness axiom to deduce the fact that between any two real numbers there is a rational and an irrational number. These are called density properties of the rationals and irrationals. We begin with the definition of the rational numbers.

Definition 4.6.1

Let $\mathbf{Q} = \{m/n \mid m, n \in \mathbf{Z}, \text{ and } n \neq 0\}$. We call $\mathbf{Q}$ the set of rational numbers. Any real number that is not rational is called irrational. •

Note that $\mathbf{Q}$ satisfies axioms A1 through A10, and so $\mathbf{Q}$ is an ordered field. The density properties depend on the following theorem.

Theorem 4.6.2 (The Archimedean Property)

Let $a, b \in \mathbf{R}$, and $a > 0$. Then there exists an $n \in \mathbf{N}$ such that $b < na$.

Proof:
Let $S = \{na \mid n \in \mathbf{N}\}$. Suppose $x \leq b$ for all $x \in S$. Then by axiom A11 $\sup(S) = c$ exists. Since $c - a < c$, there is an $n \in \mathbf{N}$ such that $na > c - a$. So,

$(n + 1)a > c$, but that is impossible since c is an upper bound of S. Thus, for some n, $na > b$. •

Theorem 4.6.3

Let $a, b \in \mathbf{R}$, and $a < b$. Then there is a rational number r such that $a < r < b$.

Proof:
First $0 < 1/(b - a)$ and $0 < 1$, so by the Archimedean Property there is an $n \in N$ such that $1/(b - a) < n1 = n$. So, $1/n < b - a$. But $0 < 1/n$, so there is an $m \in N$ such that $(1/n)m = m/n > a$. Let m_0 be the least natural number such that $a < m_0/n$. Now suppose $b \leq m_0/n$, then $\dfrac{m_0 - 1}{n} = \dfrac{m_0}{n} - \dfrac{1}{n} > b - (b - a) = a$. But that is a contradiction, so $a < m_0/n < b$. •

We shall make use of the following lemma whose proof is left as an exercise. Recall that a lemma is a theorem that is not usually important in itself, but is used in the proof of a more important theorem.

Lemma 4.6.4

Let $a, b, c \in \mathbf{R}$, $c > 0$, and $a < b$. Then there exists an $m \in N$ such that $a < a + (c/m) < b$, and there exists an $m \in N$ such that $a < b - (c/m) < b$.

We now use axiom A11 to show that every positive real number has a unique positive nth root.

Definition 4.6.5

Let $n \in N$, $a \in \mathbf{R}$, and $a > 0$. If there is a real $b > 0$ such that $b^n = a$ we say that b is an nth root of a. •

Theorem 4.6.6

Suppose $a > 0$, and $n \in N$. Then there exists a unique $b > 0$ such that $b^n = a$.

Proof:
Let $S = \{x \mid x^n \leq a\}$. Clearly $0 \in S$, so S is not empty. Also S has an upper bound of 1 if $a \leq 1$, or a if $a > 1$ (Why?). Then S has a supremum, say b. We claim that $b^n = a$. Suppose not, then $b^n < a$, or $b^n > a$. Suppose $b^n < a$. By the Binomial Theorem we have

$$(b + \frac{1}{m})^n = b^n + C_{n,1}b^{n-1}(1/m)^1 + \ldots + (1/m)^n$$
$$\leq b^n + K/m.$$

But $b^n < a$, and K is some real number, so by Lemma 4.6.4 we can choose m so that $b^n < (b + \frac{1}{m})^n < a$. But then b is not the sup of S.

The case $a < b^n$ is similar and is left to the reader. •

We use the usual notation $\sqrt[n]{a}$ or $a^{1/n}$ for nth roots.

Definition 4.6.7

Let $a > 0$, and $m, n \in \mathbf{N}$. We define $a^{m/n} = (a^{1/n})^m$. •

We can now prove the usual rules of fractional powers.

Theorem 4.6.8

Let $a > 0$, and $m, n, p, q \in \mathbf{N}$. Then,

1) $a^{(m/n) + (p/q)} = a^{m/n}a^{p/q}$,
2) $(a^{1/n})^{1/q} = a^{1/nq}$, and
3) $(a^{m/n})^{p/q} = a^{mp/nq}$.

Proof:
We have $(m/n) + (p/q) = (mq + np)/nq$, so

$$a^{(m/n) + (p/q)} = a^{(mq+np)/nq} = (a^{1/nq})^{(mq + np)} = (a^{1/nq})^{mq}(a^{1/nq})^{np} = a^{m/n}a^{p/q}.$$

Parts 2) and 3) are left to the reader. •

We have not proved that there are any irrational numbers. We turn to this now.

There Exist Irrational Numbers

We shall prove that the square root of two is irrational.

Theorem 4.6.9

$\sqrt{2}$ is irrational.

Proof:
Suppose $\sqrt{2}$ is rational. Then $\sqrt{2} = m/n$ where m and n are in $\mathbf{N}$. We may assume that at least one of m or n is odd because by the Fundamental Theorem of Arithmetic we may factor m and n into primes and cancel 2s. By definition $2 = m^2/n^2$, so $2n^2 = m^2$. Thus, m^2 is even, so m is even. (Why? Suppose

not?!) Let $m = 2k$, then $2n^2 = 4k^2$, and so $n^2 = 2k^2$. Thus, n is even also. This is a contradiction. •

We can now show that the irrational numbers are dense.

Theorem 4.6.10

Let $a < b$. Then there is an irrational number x such the $a < x < b$.

Proof:

$\frac{a}{\sqrt{2}} < \frac{b}{\sqrt{2}}$. So there is a rational number r such that

$\frac{a}{\sqrt{2}} < r < \frac{b}{\sqrt{2}}$. But then $a < \sqrt{2}\, r < b$, and $\sqrt{2}\, r$ is irrational. (Why?) •

We shall prove the existence of irrational numbers in a very different way in the next section.

We conclude this section with a discussion of the decimal representation of real numbers.

Decimal Representations

If $a \geq 0$ is any real number then there is a unique integer $a_0 \geq 0$ such that $a_0 \leq a < a_0 + 1$. Thus, a_0 is the whole number part of the decimal representation. Since $a_0 \leq a < a_0 + 1$, then $a_0 + k/10 \leq a < a_0 + (k + 1)/10$ for a unique $k = 0, 1, 2, 3, 4, 5, 6, 7, 8$, or 9. Call this k, a_1. Now since

$$a_0 + (a_1/10) \leq a < a_0 + ((a_1 + 1)/10)$$

there is a unique a_2 such that

$$a_0 + (a_1/10) + (a_2/10^2) \leq a < a_0 + (a_1/10) + ((a_2 + 1)/10^2).$$

This algorithm defines the decimal representation

$$a_0.a_1a_2 \ldots a_n \ldots .$$

We conclude that to each real number there corresponds a unique decimal representation satisfying the above criteria.

On the other hand given a decimal expansion, $b_0.b_1b_2 \ldots b_n \ldots$, one may ask if it is the decimal expansion of a unique real number obtained from the above algorithm? It is, provided we agree to "correct up" decimals that end in a string of 9s.

To see this set

$$x_n = b_0 + (b_1/10) + (b_2/10^2) + \ldots + (b_n/10^n), \text{ and}$$

$S = \{x_n \mid n \in \mathbf{N}\}$. Then $b_0 + 1$ is an upper bound for S (Why? See Exercise 6). Therefore S has a sup, say a. We also leave it to the reader to prove that if $b_0.b_1b_2 \ldots b_n \ldots$ is all 9s beyond say b_k, then the decimal $b_0.b_1b_2 \ldots b_n \ldots$ and $b_0.b_1b_2 \ldots (b_k + 1)000 \ldots$ define sets with the same sup (Exercise 7). In view of this we shall rule out decimals that have repeated 9s, so we will always "correct up". Clearly, $b_0 \le a < b_0 + 1$. If $a = b_0$ then all the $b_i = 0$, and the algorithm gives $b_0.000 \ldots$ for a. If $b_0 < a < b_0 + 1$, then $b_0 + (b_1/10) \le a < b_0 + ((b_1 + 1)/10)$ (Why? See Exercise 8). If $b_0 + (b_1/10) = a$ then $b_i = 0$ for $i \ge 2$. So the algorithm gives $b_0.b_1000 \ldots$. Continuing in this way we see that the algorithm generates the decimal we started with.

To complete the correspondence between the elements of $\mathbb{R}$ and their decimal representatives we shall show that if r_1 and r_2 are in $\mathbb{R}$ then the decimal representation of the sum of r_1 and r_2 is the sum of the decimals calculated in the way we learned in school. We shall demonstrate this only for positive numbers the extension to positive and negative numbers being straightforward. We leave the corresponding assertion for products to the exercises.

Let the decimal representations of r_1 and r_2 be $a_0.a_1a_2\ldots$, and $b_0.b_1b_2\ldots$. Let

$$r_1 = a_0 + (a_1/10) + \ldots + (a_k/10^k) + R_{1k},$$

and

$$r_2 = b_0 + (b_1/10) + \ldots + (b_k/10^k) + R_{2k},$$

then

$$(r_1 + r_2) - [(a_0 + b_0) + \ldots + (a_k + b_k)/10^k] = R_{1k} + R_{2k}.$$

We rewrite

$$(a_0 + b_0) + \ldots + (a_k + b_k)/10^k \quad \text{as}$$

$$d_0 + (d_1/10) + \ldots + (d_k/10^k)$$

where $d_i = 0, 1, 2, \ldots,$ or 9 for $i \ge 1$. It is important to note that the d_i are computed by exactly the algorithm for adding decimals learned in school. Indeed, it is the same "divide by ten and carry" procedure of addition in base ten.

There are now two possibilities, either $r_1 + r_2$ is in the open interval $(c_0, c_0 + 1)$ or $r_1 + r_2 = c_0$ where c_0 is a positive integer such that $c_0 \leq r_1 + r_2 < c_0 + 1$. If $r_1 + r_2$ is in the open interval $(c_0, c_0 + 1)$ then for k large enough

$$d_0 + (d_1/10) + \ldots + (d_k/10^k)$$

is in $(c_0, c_0 + 1)$ (Why?), and so $d_0 = c_0$. If $r_1 + r_2 = c_0$ then $d_0 = c_0$ or $c_0 - 1$, and the rest of the d_i must be 0's or 9's. If the latter, correcting up gives the decimal representation of $r_1 + r_2$. Now return to the first case and consider c_1 in the decimal representation. We have

$$c_0 + (c_1/10) \leq r_1 + r_2 < c_0 + (c_1 + 1/10).$$

If we have strict inequality on the left we may again choose k large enough that $d_0 + (d_1/10) + \ldots + (d_k/10^k)$ is in the interval. In this case d_0 and d_1 agree with c_0 and c_1. If we repeat the argument and all the inequalities on the left are strict then $d_0.d_1d_2 \ldots$ is the decimal representation of $r_1 + r_2$. If a left inequality is an equality at any stage, say k, then

$$d_0.d_1 \ldots d_k000\ldots \text{ or } d_0.d_1 \ldots (c_k - 1)999\ldots$$

is the decimal sum. The first is the decimal representation of $r_1 + r_2$, and, if we correct up, the second is also.

Once we have the result for products, we see that the structure $(\mathbb{R},+,\times)$ can be thought of as the familiar addition and multiplication of decimal expressions. In this sense there is essentially only one set $\mathbb{R}$ together with two binary operations satisfying axioms A1 through A11. We shall have more to say about this kind of uniqueness in Section 5.4.

Exercises for Section 4.6

1. Prove Lemma 4.6.4.

2. Complete the proof of Theorem 4.6.8

3. For negative fractional powers define $a^{-m/n} = \dfrac{1}{a^{m/n}}$. Prove the rules of powers of Theorem 4.6.8 for all rational numbers.

4. Without using Theorem 4.6.6, prove that 3 has a square root.

5. Prove that $1 + r + r^2 + \ldots + r^n = (1 - r^{n+1})/(1 - r)$, $r \neq 1$.

6. Prove that $b_0 + 1$ is an upper bound for $S = \{x_n \mid n \in \mathbf{N}\}$ as defined in the discussion of decimal representations.

7. Prove that if $b_k \neq 9$, but $b_{k+i} = 9$ for all $i \geq 1$ then the sup of S defined by $n.b_1 b_2 \ldots b_k 999 \ldots$ is

$$n + (b_1/10) + (b_2/10^2) + \ldots + ((b_k + 1)/10^k).$$

8. Let $a = \sup\{x_n \mid n \in \mathbf{N}\}$ where $x_n = b_0 + (b_1/10) + (b_2/10^2) + \ldots + (b_n/10^n)$. Prove that if $b_0 < a < b_0 + 1$, then $b_0 + (b_1/10) \leq a < b_0 + ((b_1 + 1)/10)$.

9. Prove that the decimal representation of $r_1 r_2$ is obtained by computing the product of the decimal representations of r_1 and r_2 by modeling your argument on the one for sums above.

4.7 Countable and Uncountable Sets

In Section 4.5 we introduced the distinction between finite and infinite sets. In this section we introduce another distinction between the sizes of sets.

Definition 4.7.1

Let S be a set. We say S is countable if and only if S is finite or there is a bijection $f : \mathbf{N} \to S$. If S is not countable we say S is uncountable. If S is countable but not finite we say S is denumerable. •

As we shall see, denumerability is the lowest order of infinity in the sense that any infinite set contains a denumerable subset.

We shall prove that every subset of a countable set is countable. To this end we prove the following lemma.

Lemma 4.7.2

Every subset of **N** is countable.

Proof:
Let S be a subset of **N**. If S is finite we are done, so assume that S is infinite. We inductively define a bijection $f : \mathbf{N} \to S$. Since $S \neq \phi$, we may choose $f(1)$

to be the least element in S. Since $S - \{f(1), f(2), \ldots f(n)\}$ is not empty it has a least element which we denote by $f(n+1)$. Thus $f: \mathbf{N} \to S$ is defined by induction. Note that $f(n) < f(n+1)$ for all n, so if $k_1 < k_2$ then $f(k_1) < f(k_2)$. Therefore, f is one to one.

If $\text{Ran}(f) \neq S$, let $m \in S - \text{Ran}(f)$. Then $f(k) < m$ for all k in $\mathbf{N}$ (induction?). But, $1 \leq f(1)$, and if $k \leq f(k)$ then $f(k+1) > f(k) \geq k$, so $f(k+1) \geq k+1$. So $k \leq f(k)$ for all k in $\mathbf{N}$. But then $f(m) \geq m$, and $f(m) < m$. This is impossible, so f is onto. •

Corollary 4.7.3

Every subset of a countable set is countable.

Proof:
Let S be countable, and $T \subset S$. If T is finite we are done. Suppose T is infinite. Then S is infinite, and there is a bijection $f: \mathbf{N} \to S$. Recall that f^* is the inverse image map. Thus, $f^*(T)$ is countable and not finite. Let $g: \mathbf{N} \to f^*(T)$ be a bijection, and $h: f^*(T) \to T$ be defined by $h(x) = f^{-1}(x)$. Then h is a bijection, and so $h \circ g: \mathbf{N} \to T$ is a bijection, so T is countable. •

Theorem 4.7.4

Let A be countable and $f: A \to B$ a function. Then $\text{Ran}(f)$ is countable.

Proof:
Let $g: A \to \text{Ran}(f)$ be defined by $g(x) = f(x)$ for all x in A. Then g is a surjection, and so it has a right inverse $h: \text{Ran}(f) \to A$. But then h has a left inverse, and so h is an injection. Define $k: \text{Ran}(f) \to \text{Ran}(h)$ by $k(x) = h(x)$ for all x in $\text{Ran}(f)$. Then k is a bijection. Now $\text{Ran}(h) \subset A$, so $\text{Ran}(h)$ is countable. But then $\text{Ran}(f)$ is countable (Why?). •

We have seen that a finite union of finite sets is finite (Exercise 6, Section 4.5). We now prove the corresponding theorem for countable sets.

Theorem 4.7.5

The union of a countable family of countable sets is countable.

Proof:
Let $\mathfrak{F}$ be a countable family. In order that the proof accommodate all cases, we make the following changes. If $\mathfrak{F}$ is finite add to $\mathfrak{F}$ the sets $\{i\}$ for each $i \in \mathbf{N}$ to obtain the denumerable family $\mathfrak{F}^*$. Let $A \in \mathfrak{F}^*$; replace A by $A^* = A \cup \mathbf{N}$ to get $\mathfrak{F}^{**}$. But, A^* is denumerable (Why? Exercise 3), and so $\mathfrak{F}^{**}$ is a denumerable family of denumerable sets. Note that $\bigcup \mathfrak{F} \subset \bigcup \mathfrak{F}^{**}$, so if we can

show that $\bigcup \mathfrak{F}^{**}$ is countable we are done. Since $\mathfrak{F}^{**}$ is denumerable we can denote the sets in $\mathfrak{F}^{**}$ by B_i where $i \in \mathbf{N}$. Since each B_i is denumerable there is a bijection $f_i : \mathbf{N} \to B_i$. Define $a_{ij} = f_i(j)$. Define $g : \mathbf{N} \to \bigcup \mathfrak{F}^{**}$ as follows. Let $g(1) = a_{11}, g(2) = a_{12}, g(3) = a_{21}, g(4) = a_{31}, g(5) = a_{22}$, etc. We are assigning a natural number to each position in the infinite matrix (a_{ij}) as shown.

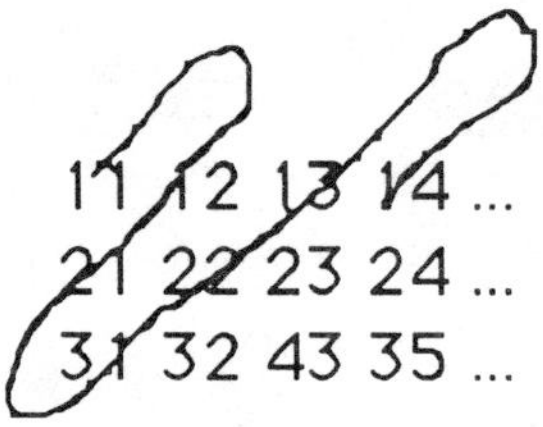

Naturally, this is called a diagonal process. We see that every entry in the matrix will be reached eventually, and so g is onto. But then by Theorem 4.7.4, $\bigcup \mathfrak{F}^{**}$ is countable, and so $\bigcup \mathfrak{F}$ is countable. •

We now give some examples of countable sets.

Example

Z is countable.

Proof:
Let $\mathbf{Z}^- = \{k \mid -k \in \mathbf{N}\}$. Then $f : \mathbf{N} \to \mathbf{Z}^-$ defined by $f(n) = -n$ is clearly a bijection, so $\mathbf{Z}^-$ is countable. But then $\mathbf{Z} = \mathbf{Z}^- \cup \{0\} \cup \mathbf{N}$ is a countable union of countable sets, hence it is countable. •

Example

The set **Q** of rational numbers is countable.

Proof:
We use the diagonal process again. Let $f : \mathbf{N} \to \mathbf{Q}^+$ be defined by $f(1) = 1/1$, $f(2) = 2/1, f(3) = 1/2, f(4) = 1/3, f(5) = 2/2$, etc. Here is how it looks.

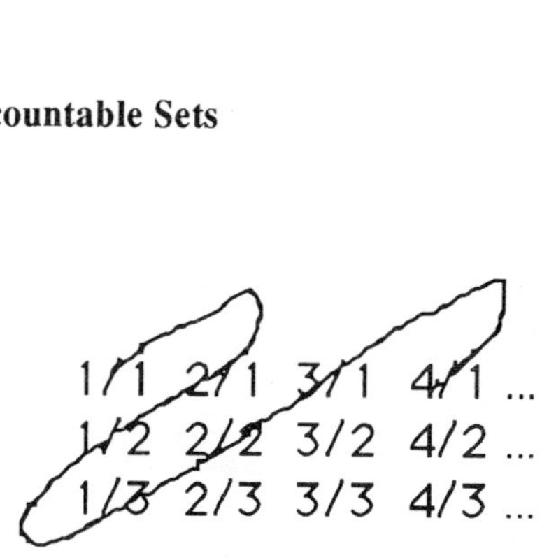

Again $f: \mathbf{N} \to \mathbf{Q}^+$ is onto, and so $\mathbf{Q}^+$ is countable. Thus $\mathbf{Q}$ is countable also (Why?). •

We now show that there are uncountable sets. The method used in the next proof is called Cantor's diagonal method after G. Cantor (1845–1918), the father of set theory.

Theorem 4.7.6

The interval [0,1) is uncountable.

Proof:
We can represent each number x in [0,1) uniquely as a decimal of the form $x = 0.a_1a_2 \ldots a_k \ldots$, where $a_k = 0, 1, 2, 3, 4, 5, 6, 7, 8,$ or 9, and the representation does not end in 9s.

Suppose [0,1) is countable, then we can list the numbers as x_1, x_2, x_3, etc. Thus we can form the array:

$$\begin{aligned} x_1 &= 0.a_{11}a_{12}a_{13} \ldots \\ x_2 &= 0.a_{11}a_{12}a_{13} \ldots \\ &\cdot \\ x_n &= 0.a_{n1}a_{n2}a_{n3} \ldots a_{nn} \ldots \\ x_{n+1} &= \text{, and so on.} \end{aligned}$$

Now choose b_i such that $b_i \neq 9$, and $b_i \neq a_{ii}$. Then there is a unique real number y in [0,1) with $0.b_1b_2b_3 \ldots$ as its decimal representation. So, since we assumed [0,1) was countable, $0.b_1b_2b_3 \ldots = 0.a_{n1}a_{n2}a_{n3} \ldots a_{nn} \ldots$ for some n. So $b_n = a_{nn}$, and $b_n \neq a_{nn}$. Thus [0,1) cannot be countable. •

Note that there must be irrational numbers since, if not, [0,1) would have to be countable.

We conclude this section by showing that given any set A there is a set B which is strictly larger in the following sense. There is an injection of A into B but there is no injection of B into A. Our candidate for B is $\mathcal{P}(A)$ – the power set of A. We can show there is no injection of $\mathcal{P}(A)$ into A by showing there is no map of A onto $\mathcal{P}(A)$.

Theorem 4.7.7

Let A be a set. Then there is no map of A onto $\mathcal{P}(A)$.

Proof:
Let $f : A \to \mathcal{P}(A)$ be a surjection. Define $S \subset A$ by $S = \{x \mid x \notin f(x)\}$. Since f is onto there is an x_0 in A such that $f(x_0) = S$. If $x_0 \in S$ then $x_0 \notin f(x_0)$, and so $x_0 \notin S$. That is a contradiction. If $x_0 \notin S$, then $x_0 \in f(x_0)$, and so $x_0 \in S$. Again this is a contradiction. So f cannot be onto. •

Exercises for Section 4.7

1. Prove that if $x \notin \mathbf{N}$ then $\mathbf{N} \cup \{x\}$ is denumerable by exhibiting a bijection from $\mathbf{N}$ to $\mathbf{N} \cup \{x\}$.

2. Prove that if A is finite and $\mathbf{N} \cap A = \phi$ then $\mathbf{N} \cup A$ is denumerable by exhibiting a bijection from $\mathbf{N}$ to $\mathbf{N} \cup A$.

3. Prove that a) if A is denumerable and $\mathbf{N} \cap A = \phi$ then $\mathbf{N} \cup A$ is denumerable by exhibiting a bijection from $\mathbf{N}$ to $\mathbf{N} \cup A$, and b) $\mathbf{N} \cup A$ is denumerable even if $\mathbf{N} \cap A \neq \phi$

4. Prove that the open interval (0,1) is uncountable.

5. Prove that there is a bijection from (0,1) to $\mathbb{R}$.

6. Is $\mathcal{P}(\mathbf{N})$ uncountable? Why? Is there a bijection from $\mathcal{P}(\mathbf{N})$ to [0,1)?

References

Robert G. Bartle and Donald R. Sherbert, *Introduction to Real Analysis* (John Wiley & Sons Inc. 1982)
Oscar Zariski and Pierre Samuel, *Commutative Algebra* (D.Van Nostrand, Princeton, New Jersey, 1963)

Chapter 5

Defining Mathematical Structures

In the Foreword to the Reader we remarked that the structures of modern mathematics have multiplied into a veritable zoo. To name a few, we have groups, rings, fields, vector spaces, topological spaces, etc. In this chapter we show, by means of three examples, how such structures are defined. In addition, we shall address ourselves to the questions of the existence and uniqueness of mathematical structures.

5.1 Groups

In this section we shall arrive at the axioms for the very important concept of a group by examining several different systems and abstracting from them a common structure.

Consider the set of integers with the operation of addition by itself. Note that there is an "identity element", 0, such that for any $x \in \mathbf{Z}$, $x + 0 = 0 + x = x$. Also note that each element x has an "additive inverse", $-x$, such that $x + (-x) = 0$. The same two properties hold for the rational and real numbers under addition.

Consider $(\mathbf{Z}_n, \oplus_n)$, the integers modulo n with addition modulo n. Again we have an identity and an inverse under the operation for each element.

Consider the nonzero rational or real numbers under the operation of multiplication. The multiplicative identity is 1, and the inverse of x is $1/x$.

Notice that, in these examples and the ones that follow, the operation is associative. That is, if "$*$" is the operation, then $a * (b * c) = (a * b) * c$ for all a, b, and c.

Now consider $(\mathbf{Z}_5 - \{[0]\}, \otimes_5)$. We have $\mathbf{Z}_5 - \{[0]\} = \{[1], [2], [3], [4]\}$. Notice that this set is closed under the operation $\otimes_5$. That is to say that when the operation is applied to pairs elements of this set the result is still in the set. Clearly, [1] is the identity. The inverses of [1], [2], [3], and [4] under this operation are [1], [3], [2], and [4] respectively. For example $[3] \otimes_5 [2] = [6] = [1]$. We leave it as an exercise to investigate the situation for other $(\mathbf{Z}_n - \{[0]\}, \otimes_n)$.

We now give a very different example. Let $G = \{e, r_1, r_2, f_1, f_2, f_3\}$. The following table defines a binary operation $\square$ on G.

$\square$	e	r_1	r_2	f_1	f_2	f_3
e	e	r_1	r_2	f_1	f_2	f_3
r_1	r_1	r_2	e	f_2	f_3	f_1
r_2	r_2	e	r_1	f_3	f_1	f_2
f_1	f_1	f_3	f_2	e	r_2	r_1
f_2	f_2	f_1	f_3	r_1	e	r_2
f_3	f_3	f_2	f_1	r_2	r_1	e

For example, to compute $r_1 \square f_1$ find r_1 in the left column and f_1 in the top row. The intersection of the row of r_1 and the column of f_1 is f_2, so $r_1 \square f_1 = f_2$.

This table was derived from the following concrete situation. Consider an equilateral triangle made of cardboard that fits into an equilateral socket like a child's shape matching game. We consider all the different ways in which the triangle can be picked up and replaced in the socket in a different position. To keep track of the possibilities number the vertices of the socket and triangle 1, 2, and 3 in clockwise order. At the start the triangle and socket numbers correspond as follows, $1 \to 1$, $2 \to 2$, and $3 \to 3$. The symbol e stands for the motion of the triangle that makes no change in the position. The motion r_1 is a one third clockwise rotation. After r_1 the correspondences are $1 \to 2$, $2 \to 3$, and $3 \to 1$. Motion r_2 is a two thirds rotation. The motion f_i is the reflection about the axis through the ith vertex and the opposite side. For example, f_1 gives the correspondence $1 \to 1$, $2 \to 3$, and $3 \to 2$. The table summarizes the effect of following one motion by another.

If we examine the table it becomes clear that there is an identity for this operation. It is e. Further examination of the table shows that each element has an inverse that carries the triangle back to the starting position. Moreover, it does not matter in which order the motion and its inverse are carried out. For example, $r_1 \square r_2 = r_2 \square r_1 = e$.

These examples motivate the following definition.

Definition 5.1.1

A group $(G,*)$ is a set G together with a binary operation $*$ such that :

G1) For all a, b, and c in G, $a * (b * c) = (a * b) * c$.

G2) There exists an element e in G such that for all x in G,

$$e * x = x * e = x.$$

G3) For each x in G there is a y in G such that

$$x * y = y * x = e. \bullet$$

We call G1) the associative law. In G2) e is called an identity. In G3) y is called an inverse for x. In the first exercises on groups the reader will be asked to prove that e and y are unique, so we can speak of the identity e and the inverse y of x.

The collection of groups is very large. It is certainly infinite since it contains all the groups $(\mathbf{Z}_n, \oplus_n)$. Indeed, the collection of all groups is too large to be a set – it is a class. In view of this, if we wish to establish truths about all groups, we cannot check each group, so we must employ the method of deductive proof. Reflect for a moment on the enormous power of deduction from the axioms G1, G2, and G3. Consider a theorem deduced from the axioms. By the nature of deduction if the axioms are true then the theorem must be true. Thus, this theorem must be true of every group in the class of groups.

Definitions that Narrow the Class

Zoologists often narrow their study of all animals to, say, the great cats. Similarly, the class of all groups can be broken up into subclasses. Here are two important subclasses defined by adding to our axioms.

Note that we have omitted the operation symbol in the next definition. Since there is only one operation this does not lead to any confusion.

Definition 5.1.2

A group G is said to be abelian if and only if for all a and b in G $ab = ba$. •

All the examples given leading up to the definition of a group were abelian except the group of motions of the equilateral triangle. This latter group we shall denote by S_3 from now on.

Definition 5.1.3

A group G is said to be finite if and only if G is a finite set. If G is finite, the number of elements in G is called the order of G and is denoted by $|G|$. •

For example the groups $(\mathbf{Z}_n, \oplus_n)$ and S_3 are finite. Note that $|\mathbf{Z}_n| = n$, and the order of S_3 is six.

We now have the subclasses of abelian groups and finite groups. Of course, we can combine these definitions and consider the subclass of finite abelian groups.

Definitions Inside Groups

It often happens that we wish to single out or highlight a property of elements or structures within a group. An example of this type of thing is the notion of a prime number in the integers. After introducing some useful notation we shall give some examples of definition within a group.

If G is a group then if $a \in G$ and n is a positive integer we write

$$a^n = aa \ldots a \ \{n \text{ factors}\}.$$

You may use definition by induction here if you want to be very rigorous.

Recall that the identity and the inverse of an element are unique (see exercises 1 and 2 below). So if $a \in G$, we write a^{-1} for the inverse of a. It is natural to set $a^{-n} = (a^{-1})^n$. We also set $a^0 = e$.

Definition 5.1.4

Let $a \in G$. If there is a positive integer n such that $a^n = e$, the least such positive integer is called the order of a, and it is denoted by $|a|$. If no such integer exists we say the order of a is infinite. •

Example

Let us find the order of every element of $\mathbf{Z}_4$ where the operation is addition modulo 4. Obviously, $|[0]| = 1$. In this group a^k means $a \oplus_4 a \oplus_4 \ldots \oplus_4 a$ where there are k summands. So, $[1]^4 = [0]$, $[2]^2 = [0]$, and $[3]^4 = [0]$. Therefore, since no lower powers give the identity $[0]$, the orders are $|[1]| = |[3]| = 4$, and $|[2]| = 2$. •

Definition 5.1.5

Let $H \subset G$. We say H is a subgroup of G if and only if H together with the operation on G is a group. •

It is important to note that this definition requires that, if H is a subgroup of G, the binary operation on G restricted to H must be a binary operation on H. This means that if a and b are in H then ab is in H also. If the latter holds we say the operation is closed on H.

Example

Return to the group of motions of the equilateral triangle S_3. Let $H = \{e, r_1, r_2\}$. A check of the operation table for S_3 shows that the operation is closed on H. Moreover, H contains the identity, and each element of H has an inverse in H. •

The following exercises provide a very short excursion into the theory of groups. The exercises should be done in order, for some exercises depend on predecessors.

Exercises for Section 5.1

1. Show that the identity in a group is unique. Hint, suppose not.

2. Show that the inverse of a group element is unique.

3. Let G be a group. Let a, b, and c be elements of G. Prove that if $ab = ac$ then $b = c$.

4. Let G be a group, and $a \in G$. Prove that for any integers m and n, $a^m a^n = a^{m+n}$, and $a^{mn} = (a^m)^n$.

5. Let G be an abelian group, and let $a, b \in G$. Prove that for any integer n, $(ab)^n = a^n b^n$. Is this true for nonabelian groups? Prove or give a counter-example.

6. Let G be a group, and let $a, b \in G$. Prove that $(ab)^{-1} = b^{-1}a^{-1}$.

7. A group G is said to be cyclic if there is an element a in G such that $G = \{a^k \mid k \in \mathbf{Z}\}$. We shall say that a generates G. Prove that all cyclic groups are abelian. Are all abelian groups cyclic?

8. Let G be a group. Suppose that for any $a \in G$, $a^2 = e$. Prove that G is abelian.

9. Let H be a subgroup of the group G. Prove that the identity in H is the same as that in G. Also prove that the inverse of an element in H is the same as the inverse of that element regarded as a member of G.

10. Let G be a group. Let $H \subset G$. Prove that H is a subgroup of G if and only if $H \neq \phi$, and for all a and $b \in H$, $ab \in H$, and $a^{-1} \in H$. In words, H is a subgroup of G if and only if H is not empty, and H is closed under the operation and taking inverses.

11. Let H and K be subgroups of the group G. Prove that $H \cap K$ is a subgroup

of G. What about $H \cup K$?

12. Let G be a group, and $a \in G$. Prove that if $a^m = e$ then the order of a divides m. Hint, suppose not and use the division algorithm $m = nq + r$ where $n = |a|$. Also prove that $a^m = a^r$.

13. Let G be a group. Let $a \in G$. Let $<a> = \{a^k \mid k \in \mathbf{Z}\}$. Prove that $<a>$ is a subgroup of G. We call $<a>$ the cyclic subgroup of G generated by a. Prove that $|<a>| = |a|$. Hint: If $|a|$ is infinite consider the map $k \to a^k$, and if $|a| = n$ consider $\{e, a, a^2, \ldots, a^{n-1}\}$.

14. Let H be a subgroup of the group G. Define for each $x \in G$, $xH = \{xh \mid h \in H\}$. Prove that for any x and y in G, $xH = yH$ or $xH \cap yH = \phi$.

15. Observe that in problem 14) $\bigcup \{xH \mid x \in G\} = G$. Thus, the family $\{xH \mid x \in G\}$ partitions G. What is the corresponding equivalence relation? The sets xH are called left cosets of H in G.

16. Let H be a subgroup of G. For any $x \in G$, prove that there is a bijection from H to xH. Thus, all left cosets have the same number of elements in them.

17. Let G be a finite group, and H a subgroup. Prove that $|H|$ divides $|G|$. This is the famous theorem of Lagrange.

18. Let G be a finite group, and $a \in G$. Prove that $|a|$ divides the order of G.

19. If $|G| = 5$ what kind of a group is G? Can you generalize?

5.2 Consistency

In the course of this book we have, so far, defined two mathematical structures using the theory of sets as a base. We have the definition of the real number system in Chapter 4 and the concept of a group just defined. In both cases the definition of the structure amounted to the formulation of some propositions in the language of set theory. We call these propositions axioms. Any structure for which the axioms are true is an example of the structure defined by those axioms. For example, the set of motions S_3 of an equilateral triangle together with the operation of following one motion by another is an example of a group structure. Is it possible to define a mathematical structure for which no examples exist?

Suppose the axioms of our definition imply a contradiction. Then there could be no example of the structure defined by those axioms, since, in that case, the axioms would be true and could not imply a false conclusion. Therefore, the first requirement of a definition of a mathematical structure is that the axioms employed not imply a contradiction.

Definition 5.2.1

We shall say a set of axioms is consistent if and only if they do not imply a contradiction. We say they are inconsistent otherwise. •

This is all very reasonable, but how do you test a set of axioms to see if it is consistent? One way is to find an example of the structure defined by the axioms. For this example the axioms are true, so the axioms could not imply anything false. Therefore a contradiction could not be deduced from the axioms.

Clearly, the axioms for a group are consistent because we have a concrete example of a group – namely, S_3. How about the 11 axioms for the real number system? In any example of a real number system the set $\mathbb{R}$ will have to be infinite. If you contend that infinite sets cannot exist in the real world, then it is not possible to establish the consistency of the axioms for the real numbers as we did for the axioms of a group. On the other hand, if you contend that the set of infinite decimals exists in the world of concepts and addition and multiplication in this set is defined in the usual way, then you have an example of the real number structure. In this case you would conclude that the real number axioms are consistent.

Clearly the question of the consistency of the real number axioms is problematical. It may come as a surprise to find the existence of the most important number system called into question. Yet, while practically no one doubts the consistency of the real number system, no one has been able to provide an unimpeachable proof of consistency. It is this author's opinion that even if the real number axioms were to prove inconsistent neither the world nor the real numbers would come to an end. Instead, the axioms would be reformulated to avoid the contradiction, and a deeper and more subtle real number system would be developed.

Exercise for Section 5.2

Define a tesop to be a set S together with a relation $\neg$ on S such that a) for all x in S $x \neg x$, b) for all x and y in S if $x \neg y$ and $y \neg x$ then $x = y$, and c) for all x, y, and z in S if $x \neg y$ and $y \neg z$ then $x \neg z$. Find two examples of tesops, one where S is infinite and one where S is finite. What do you say about the consistency of the tesop axioms?

5.3 Finite Probability Spaces

In this section we show how the concept of a finite probability space is defined. This will provide an example of the definition of a mathematical structure that differs from the purely "algebraic" structures of the real number system and groups.

In the game of craps two dice are thrown and the sum of the numbers that face up is crucial in the play of the game. Therefore, the probabilities of the occurrence of the possible sums is of central importance in the betting. To analyze these probabilities we first consider all the possible outcomes. The outcomes are pairs of numbers (i,j) where i is the up facing number on die 1 and j is that on die 2. Thus, $S = \{(i,j) \mid i, j = 1, 2, \ldots, \text{ or } 6\}$ is the set of all possible outcomes of a roll of the dice. We call S the sample space. Here, S has 36 elements.

If the dice are fair we expect that the ratio of the number of occurrences of a certain pair (i,j) to the total number of rolls will tend to 1/36 as the number of rolls increases. Since the outcomes that sum to 7 are

$$\{(1,6), (6,1), (2,5), (5,2), (3,4), (4,3)\},$$

and there are six of them, we expect the ratio of rolls that sum to 7 to the total number of rolls to tend to 6/36 or 1/6 as the number of rolls tends to infinity. We say the probability of rolling 7 is 1/6. We could bet on many other events. For example, we could bet on the occurrence of a "2". This event is described by the subset of S,

$$\{(2,1), (2,2), (2,3), (2,4), (2,5), (2,6), (1,2), (3,2), (4,2), (5,2), (6,2)\}.$$

We would expect the probability to be 11/36. In short, events are described by subsets of S, and each subset is assigned a probability which is a number between 0 and 1.

What is the probability of rolling 7 or a double 1 or double 2? The event is

$$\{(1,6), (6,1), (2,5), (5,2), (3,4), (4,3), (1,1), (2,2)\}.$$

So the probability is 8/36 or 2/9. But this event is the union of $A = \{(1,6), (6,1), (2,5), (5,2), (3,4), (4,3)\}$, and $B = \{(1,1), (2,2)\}$. The probabilities of these events are 1/6 and 1/18 respectively. Note that $1/6 + 1/18 = 2/9$. These considerations motivate the following definition of a finite probability space.

Definition 5.3.1

A finite probability space is a finite set S together with a function $P : \mathcal{P}(S) \to [0,1]$ such that the following axioms hold.

P1) $P(\phi) = 0$, and $P(S) = 1$.

P2) For any subsets A and B of S, if $A \cap B = \phi$ then

$$P(A \cup B) = P(A) + P(B).$$

We call S the sample space and any subset of S an event. We call the function P a probability measure on S (even though it is on the power set of S, $\mathcal{P}(S)$). Of course [0,1] is the closed interval from 0 to 1. Property P2) is called finite additivity. •

The following exercises, like those for groups, provide a short introduction to the development of the theory of finite probability spaces.

Exercises for Section 5.3

Throughout, S is a finite probability space and P is a probability measure on S.

1. Give two examples of finite probability spaces different from the one above.

2. If A is a subset of S, prove that $P(S - A) = 1 - P(A)$.

3. If $A_1, A_2, \ldots, A_n$ are subsets of S such that $A_i \cap A_j = \phi$ if $i \neq j$ prove that $P(A_1 \cup \ldots \cup A_n) = P(A_1) + \ldots + P(A_n)$. (induction?)

4. If A and B are subsets of S, prove that $P(A \cup B) = P(A) + P(B) - P(A \cap B)$.

5. If A and B are subsets of S, prove that $P(A \cup B) \leq P(A) + P(B)$.

6. If A and B are subsets of S, and $A \subset B$, prove that $P(A) \leq P(B)$.

7. If A and B are subsets of S, $A \subset B$, and $P(B) = 0$, prove that $P(A) = 0$.

8. Let X be a nonempty subset of S. Let $P_X : \mathcal{P}(S) \to [0,1]$ be defined by $P_X(A) = P(A \cap X)/P(X)$ if $P(X) \neq 0$. Prove that P_X is a probability measure on S.

We pause in the exercises to comment on problem 8. The probability measure P_X is called the conditional probability given X. The concept is motivated by the following. In our game of dice suppose we only consider rolls that show at least one "2". Then our new sample space consists of those occurrences that show a "2". We can call this set X. Now if we want the probability of rolling 7 in our new sample space we must find the ratio of the number of rolls that gave 7 and showed a "2" to the number of rolls that showed a "2". If n is a large number of rolls then this ratio is approximately

$$nP(A \cap X)/nP(X) = P_X(A),$$

where A is the event of rolling a sum of 7.

The critical reader may object that, since the sample space is now X, the events should be subsets of X and not S. Exercise 9) shows that this view is equivalent.

9. Let X be a nonempty subset of S, and $P(X) \neq 0$. Define $\underline{P} : \mathcal{P}(X) \to [0,1]$ by $\underline{P}(B) = P(B)/P(X)$ where B is any subset of X. Prove that $\underline{P}$ is a probability measure on X. Also prove that if A is any subset of S such that $A \cap X = B$ then $P_X(A) = \underline{P}(B)$.

In exercise 8. it could happen that $P_X(A) = P(A)$. In this case the probability of event A when no restriction is applied is the same as when we restrict our attention only to occurrences in X. We say A and X are independent events.

10. Prove that if $P(X) \neq 0$ then A and X are independent events if and only if $P(A \cap X) = P(A)P(X)$.

The condition that $P(A \cap X) = P(A)P(X)$ is usually taken as the definition of independence. This allows $P(X) = 0$. We shall adopt this definition.

11. Find two independent events in the game of craps. Of course! Now find a nontrivial pair.

12. Prove that if A and X are independent then so are $S - A$ and X.

13. Let (S_1, P_1) and (S_2, P_2) be finite probability spaces. Prove that there is exactly one probability measure $\underline{P}$ on $S_1 \times S_2$ such that if A and B are subsets of S_1 and S_2, respectively then $\underline{P}(A \times B) = P_1(A)P_2(B)$. We denote this measure by $P_1 \times P_2$. It is called the product measure.

14. Let S_1 and S_2 be finite sets. Let $\underline{P}$ be a probability measure on $S_1 \times S_2$. Prove that P_1 and P_2 defined by $P_1(A) = \underline{P}(A \times S_2)$ and $P_2(B) = \underline{P}(S_1 \times B)$ are probability measures on S_1 and S_2, respectively.

15. In exercise 14. what is necessary and sufficient in order that $\underline{P}$ be the product measure $P_1 \times P_2$?

16. Are the axioms for a finite probability space consistent?

5.4 Uniqueness of Mathematical Structures

We have infinitely many examples of groups and finite probability spaces, but how many real number systems as defined in Chapter 4 do we have? To answer this we must first decide when two examples of a structure are the same or different.

Consider the two groups $\mathbf{Z}_2 = \{[0], [1]\}$ under addition modulo 2, and $G = \{1, -1\}$ under ordinary multiplication. In one sense these groups are different, for they are, at the least, different sets. However, in a sense we shall now explain, they are the same.

Consider the map $f : \mathbf{Z}_2 \to G$ defined by $f([0]) = 1$ and $f([1]) = -1$. The map f is a bijection, and it is quickly checked that if x and y are in $\mathbf{Z}_2$ then $f(x \oplus_2 y) = f(x)f(y)$. The correspondence f preserves the group operations. As a consequence, we can use f to translate any statement about $\mathbf{Z}_2$ into a statement about G. These statements will either both be true or both false. For example, in $\mathbf{Z}_2$, [1] is the inverse of [1], but $f([1]) = -1$, so the translation is -1 is the inverse of -1. Thus there is nothing you can learn about $\mathbf{Z}_2$ that you have not automatically learned about G. In this sense $\mathbf{Z}_2$ and G are the same. We make the following definition.

Definition 5.4.1

We say two groups $(G_1,*)$ and $(G_2, \circ)$ are isomorphic if and only if there is a bijection $f : G_1 \to G_2$ such that for all x and y in G_1, $f(x * y) = f(x) \circ f(y)$. We call f an isomorphism. •

Thus, isomorphic groups are regarded as the same group. Clearly, there are infinitely many nonisomorphic groups because isomorphic groups must have the same number of elements in them.

What about examples of the real number system? We say two examples of the real number system are isomorphic if there is a bijection between them that preserves both the operations of addition and multiplication and the order relation $\leq$. Such a bijection always exists. To see this let the examples be $\mathbb{R}_1$ and $\mathbb{R}_2$. If r is in $\mathbb{R}_1$ we define $f(r)$ to be the element in $\mathbb{R}_2$ that has the same decimal expansion as r, as described in Section 4.5. But in Section 4.5 we showed that the decimal representation of the sum and product of elements of $\mathbb{R}_1$ or $\mathbb{R}_2$ is the sum and product of the decimals. But this says that f must preserve both operations. It also preserves the ordering. In view of these considerations, the real number system has, up to isomorphism, only one example. Because of this, we can speak of the real number system.

We shall conclude this section with a much simpler example of a structure that has, in essence, only one example.

The reader may have wondered why we defined the real number system directly and then defined the natural numbers as a subsystem. This seems a little like putting the cart before the horse. After all, we learn to count first and then build up from there. It is certainly possible to start with the natural numbers and construct the integers, rationals, and reals from there. The author contemplated this approach for Chapter 4, but decided against it on the grounds that the development might overwhelm the reader at that stage. Now, however, we can sketch such a development.

If we are to begin with the natural numbers how should we first define them? We lay down axioms that capture the notion of a sequence.

Definition 5.4.2

We say $(\mathbf{W}, S)$ is a whole number system if and only if $\mathbf{W}$ is a set and $S : \mathbf{W} \to \mathbf{W}$ is a function such that:

N1) There is an element 1 in $\mathbf{W}$ such that 1 is not in the range of S.

N2) For all m and n in $\mathbf{W}$, if $S(m) = S(n)$ then $m = n$.

N3) For any subset T of $\mathbf{W}$, if $1 \in T$, and for all $n \in \mathbf{W}$, if $n \in T$ then $S(n) \in T$, then $T = \mathbf{W}$. •

We say that $S(n)$ is the successor of n and n is the predecessor of $S(n)$. Note that 1 is not the successor of any element. The reader should not be surprised to find that we call N3) the induction axiom. These axioms are known as the Peano axioms after the Italian mathematician Peano. An example of a system that satisfies the Peano axioms is given by $\mathbf{N}$, the natural numbers, and S defined by $S(n) = n + 1$.

One can ask if every element except 1 has a predecessor.

Theorem 5.4.3

Every element except 1 has a predecessor.

Proof:
Let $T = \{n \in \mathbf{W} \mid n = 1 \text{ or } n = S(m) \text{ for some } m \in \mathbf{W}\}$. Trivially, $1 \in T$. Now suppose $n \in T$, then $n = S(m)$ for some m. But then $S(n) = S(S(m))$, and so $S(n) \in T$. So, $T = \mathbf{W}$. •

We now show that all whole number systems are essentially the same.

Definition 5.4.4

We shall say that two whole number systems $(\mathbf{W}_1, S_1)$ and $(\mathbf{W}_2, S_2)$ are isomorphic if and only if there is a bijection $f: \mathbf{W}_1 \to \mathbf{W}_2$ such that for all n in $\mathbf{W}_1$, $f(S_1(n)) = S_2(f(n))$. The successor relation is preserved. •

Theorem 5.4.5

All whole number systems are isomorphic.

Proof:
Let $(\mathbf{W}_1, S_1)$ and $(\mathbf{W}_2, S_2)$ be two such systems. Define a function f from $\mathbf{W}_1$ to $\mathbf{W}_2$ by $f(1_1) = 1_2$, and if $n \in \mathbf{W}_1$ and $f(n)$ has been defined then define $f(S_1(n)) = S_2(f(n))$. Clearly, by N3) f is defined for all n. Now $1 \in \text{Ran}(f)$ (we drop the subscripts on the 1's), and if $m \in \text{Ran}(f)$ then $m = f(n)$ for some n. But then $S_2(m) = S_2(f(n)) = f(S_1(n))$, and so $S_2(m) \in \text{Ran}(f)$. So f is onto.

Suppose $f(1) = f(m)$ and $m \neq 1$. Then $m = S_1(k)$ for some k. So, $1 = f(S_1(k)) = S_2(f(m))$, but that is impossible by N1), so $m = 1$.

We now prove that for all m and for all n in $\mathbf{W}_1$, if $f(m) = f(n)$ then $m = n$. Let $T = \{m \mid \text{for all } n, \text{ if } f(m) = f(n) \text{ then } m = n\}$. Clearly, $1 \in T$. Suppose $m \in T$. Consider $S_1(m)$. If $f(n) = f(S_1(m))$ then $n \neq 1$, and $n = S_1(k)$ for some k. So $f(S_1(m)) = f(S_1(k))$, and so $S_2(f(m)) = S_2(f(k))$. Then, $f(m) = f(k)$, and so $m = k$. Therefore, $S_1(m) = n$. Then, $S_1(m) \in T$. So, $T = \mathbf{W}_1$. •

The isomorphism $f: \mathbf{W} \to \mathbf{N}$ can be used as follows to define operations of addition and multiplication on $\mathbf{W}$. Set $m + n = f^{-1}(f(m) + f(n))$ and $mn = f^{-1}(f(m)f(n))$. The laws of arithmetic follow immediately from this. However, in the exercises the reader is asked to develop the laws of arithmetic entirely inside $\mathbf{W}$ using the Peano axioms only.

Exercises for Section 5.4

In exercises 1 through 8 use only the axioms for a whole number system $(\mathbf{W}, S)$.

1. For all m and n in $\mathbf{W}$, define $m + 1 = S(m)$, $m + S(n) = S(m + n)$, $m1 = m$, and $mS(n) = mn + m$. Prove that $m + n$ and mn are defined for all m and n.

2. Prove that $m + 1 = 1 + m$ for all m in **W**.

3. Prove that for all m and n in **W**,
   a) $(m + n) + 1 = m + (n + 1)$, and
   b) $1 + (m + n) = (1 + m) + n$.

4. Prove that for all m and n in **W**, $m + n = n + m$.

5. Prove that for all m, n, and k in **W**, $m + (n + k) = (m + n) + k$.

6. Prove that for all m, n, and k in **W**, $m(n + k) = mn + mk$.

7. Prove that for all m, n, and k in **W**, $m(nk) = (mn)k$.

8. Prove that $m1 = 1m$ for all m in **W**.

9. Prove that $mn = nm$ for all m and n in **W**.

In this next exercise we assume our knowledge of **N** again.

10. Let $p \in$ **N** with p a prime number. Prove that all groups of order p are isomorphic. Hint, prove that all groups of order p are cyclic. Hint, see Lagrange's Theorem.

5.5 An Outline of the Construction of the Real Numbers

We shall assume from now on that we have the natural numbers **N**. In this section we shall show how the integers, rationals, and reals can be constructed from **N**. This construction is given in outline only, and many details have been left out. However, the attentive reader will have no difficulty in supplying the missing details. There is only one exercise for this section and it is to fill in the details.

The Integers

If we already have the integers then the expression $m - n$ where m and n are natural numbers makes sense. Moreover, $(m + k) - (n + k)$ denotes the same integer for any k. However, if we only have the natural numbers at our disposal $m - n$ only makes sense if $m > n$. If $m < n$ then $m - n$ is only a pair of numbers in search of something to stand for. These considerations motivate the following construction of the integers.

Consider $\mathbf{N} \times \mathbf{N} = \{(m,n) \mid m, n \in \mathbf{N}\}$. We think of (m,n) as a typical formal difference $m - n$. But this pair or difference may not stand for anything. We now show how to construct something for (m,n) to denote. We define an equivalence relation $\sim$ on $\mathbf{N} \times \mathbf{N}$ by $(m,n) \sim (p,q)$ if and only if $m + q = p + n$. This is motivated by the fact that if we had the integers then $m - n = p - q$ if and only if $m + q = p + n$. We leave it to the reader to show that $\sim$ is an equivalence relation. The equivalence classes will be denoted by $<(m,n)>$. The set of equivalence classes is denoted by $\mathbf{Z}$.

We now show how to add and multiply the elements of $\mathbf{Z}$. Addition is defined by

$$<(m,n)> + <(p,q)> = <(m + p, n + q)>.$$

Multiplication is defined by

$$<(m,n)><(p,q)> = <(mp + nq, np + mq)>.$$

We remind the reader that we are adding and multiplying equivalence classes by using representatives, so well-definition must be checked. We check addition and leave multiplication to the reader. Let $<(a,b)> = <(c,d)>$ and $<(m,n)> = <(p,q)>$. Then, $(a,b) \sim (c,d)$ and $(m,n) \sim (p,q)$. So, $a + d = c + b$, and $m + q = p + n$. We must show that $<(a + m, b + n)> = <(c + p, d + q)>$. But, $a + m + d + q = c + p + b + n$, so we are done. With these operations on $\mathbf{Z}$ it is straightforward to verify that axioms A1 – A8 of Chapter 4 are satisfied.

For example consider the distributive law. We have

$$<(a,b)>[<(c,d)> + <(e,f)>] = <(a,b)><(c + e, d + f)>$$

$$= <(a(c + e) + b(d + f), b(c + e) + a(d + f))>$$

$$= <(ac + ae + bd + bf, bc + be + ad + af)>.$$

But

$$<(a,b)><(c,d)> + <(a,b)><(e,f)>$$

$$= <(ac + bd + ae + bf,\ bc + ad + af + be)>.$$

We leave the rest to the reader.

In $\mathbf{Z}$ we denote $<(m,m)>$ by 0. Note that $<(a,b)> + 0 = <(a,b)>$. The class $<(1 + 1,1)>$ is denoted by 1. Note that $<(a,b)>1 = <(a,b)>$.

It is most important to note that the set

$$\underline{\mathbf{N}} = \{<(n + 1,1)> \mid n \text{ is in } \mathbf{N}\}$$

is isomorphic to **N** (prove this). Put another way, $\underline{\mathbf{N}}$ is a whole number system and so $\underline{\mathbf{N}}$ and **N** are essentially the same. Thus we identify $\boldsymbol{n}$ with $<(\boldsymbol{n} + 1,1)>$. If we write $-\boldsymbol{n}$ for $<(1,\boldsymbol{n} + 1)>$ then every element of **Z** is either $\boldsymbol{n}$, 0, or $-\boldsymbol{n}$ where $\boldsymbol{n}$ is a natural number. In this way the complexities of the equivalence classes melt away, and **Z** emerges as the set of integers so familiar to us.

Finally, it is useful to note that **Z** satisfies the following. If $\boldsymbol{x}$ and $\boldsymbol{y}$ are in **Z** and $\boldsymbol{xy} = 0$, then $\boldsymbol{x} = 0$ or $\boldsymbol{y} = 0$ (prove this).

The Rationals

In learning about numbers we learn early on about fractions such as 5/2, 3/6, or 1/2. We soon learn that 3/6 and 1/2 are to be considered the same. Indeed, if we already have the rational numbers then fraction $\boldsymbol{m/n}$ of integers make sense, and $\boldsymbol{mk/nk} = \boldsymbol{m/n}$ is a simple theorem. But, if the rationals do not as yet exist, a fraction $\boldsymbol{m/n}$ is again a symbol in search of a referent.

These observations motivate the definition of the set of rational numbers **Q**. Consider

$$\mathbf{Z} \times (\mathbf{Z} - \{0\}) = \{(\boldsymbol{a},\boldsymbol{b}) \mid \boldsymbol{a}, \boldsymbol{b} \in \mathbf{Z}, \boldsymbol{b} \neq 0\}.$$

We define an equivalence relation $\approx$ on this set by $(\boldsymbol{a},\boldsymbol{b}) \approx (\boldsymbol{c},\boldsymbol{d})$ if and only if $\boldsymbol{ad} = \boldsymbol{bc}$. This is quite natural if we think of $(\boldsymbol{a},\boldsymbol{b})$ as $\boldsymbol{a/b}$, $(\boldsymbol{c},\boldsymbol{d})$ as $\boldsymbol{c/d}$, and $(\boldsymbol{a},\boldsymbol{b}) \approx (\boldsymbol{c},\boldsymbol{d})$ as identifying $\boldsymbol{a/b}$ with $\boldsymbol{c/d}$. We denote the equivalence classes by $[(\boldsymbol{a},\boldsymbol{b})]$, and denote the set of equivalence classes $\mathbf{Z} \times (\mathbf{Z} - \{0\})/\!\approx$ by **Q**.

Addition and multiplication in **Q** are defined as follows. We define $[(\boldsymbol{a},\boldsymbol{b})] + [(\boldsymbol{c},\boldsymbol{d})] = [(\boldsymbol{ad} + \boldsymbol{bc}, \boldsymbol{bd})]$, and $[(\boldsymbol{a},\boldsymbol{b})][(\boldsymbol{c},\boldsymbol{d})] = [(\boldsymbol{ac},\boldsymbol{bd})]$. These definitions are quite natural if one recalls the way in which fractions are added and multiplied. The reader should carefully note that we are using the names of two equivalence classes to define a third. So it must be asked if the definitions are well-defined. They are, and the proof is left to the reader. Also left to the reader is the proof that **Q** together with these operations is an ordered field satisfying axioms A1 – A10 of Chapter 4. We call **Q** the rational number system.

The set $\{[(\boldsymbol{n},1)] \mid \boldsymbol{n} \in \mathbf{Z}\}$ is an isomorphic copy of the integers. That is, there is a bijection from this set to **Z** that preserves the operations of addition and multiplication. So we may think of the integers as a subset of the rationals.

The Reals

In Chapter 4 it was shown that the real numbers can only be "cut" in two ways. They are $\mathbb{R} = (-\infty,\boldsymbol{r}) \cup [\boldsymbol{r},\infty)$, and $\mathbb{R} = (-\infty,\boldsymbol{r}] \cup (\boldsymbol{r},\infty)$. This is not the case with the rational numbers. For example the cut

$$\mathrm{L} = \{q \in \mathrm{Q} \mid q^2 < 2\} \text{ and } \mathrm{R} = \{q \in \mathrm{Q} \mid q^2 > 2\}$$

is not of those types.

This fact motivates the definition of the real numbers as the set of all cuts in **Q**. A cut in **Q** is a pair of subsets of **Q**, $[A,B]$, such that A and B are not empty, if $a \in A$ and $b \in B$ then $a < b$, and $A \cup B = \mathbf{Q}$. If a cut is of the form $[(-\infty,r),[r,\infty)]$, where r is a rational number, then we also have the cut $[(-\infty,r],(r,\infty)]$. We identify such cuts and denote the identification by $[r]$ and call it the rational cut at r. We call all other cuts irrational. A cut is determined by its left set. It is convenient to define a cut in terms of a single set. We redefine a cut as a nonempty set A such that $A \neq \mathbf{Q}$, and 1) if $s \in A$ and $q \in \mathbf{Q}$, and $q < s$ then $q \in A$, and 2) if $s \in A$ there is a $t \in A$ such that $s < t$. Thus our only choice for a rational cut is $(-\infty,r)$. The set of real numbers, $\mathbb{R}$, is defined to be the set of all rational and irrational cuts of **Q**.

We now show how to add cuts. Let A_1 and A_2 be cuts. Define

$$A_1 + A_2 = \{r_1 + r_2 \mid r_1 \in A_1 \text{ and } r_2 \in A_2\}.$$

Then $A_1 + A_2$ is a cut which we call the sum of A_1 and A_2.

To define the product of cuts we first define the product of "positive" cuts. A positive cut is a cut A in which A contains at least one positive rational. Denote the set of positive cuts by $\mathbb{R}^+$. Suppose A_1 and A_2 both contain positive numbers. For any sets of rational numbers S and T define

$$ST = \{st \mid s \in S \text{ and } t \in T\}.$$

We now define $A_1 \times A_2$ by

$$A_1 \times A_2 = (A_1 - (-\infty,0])(A_2 - (-\infty,0]) \cup (-\infty,0].$$

To make sense out of this definition consider the product of the cuts $(-\infty,2)$ and $(-\infty,3)$ to see that the definition gives the cut $(-\infty, 6)$, as we would hope it would. We leave it to the reader to prove that the above definition does define a cut.

If S is any set of rationals define S^- by $S^- = \{-s \mid s \in S\}$. Then if A is a cut it is rational or irrational. If it is rational $A = (-\infty,r)$. We define $-A = (-\infty,-r)$. $-A$ is the rational cut at $-r$. If A is irrational then $(\mathbf{Q} - A)^-$ is also a cut and we denote it by $-A$. Note that if A is a positive cut then $-A$ is not. If a cut is not positive nor the cut $[0]$ we say it is negative. If A is negative then $-A$ is positive. If $r = A$ we set $-r = -A$. We define the product of non-positive cuts as might be expected. For any cut r we define $[0]r = r[0] = [0]$. If r is a negative cut and s is a positive cut we define $rs = -(-r)s$. If r and s are negative set $rs = (-r)(-s)$.

With these operations it can be shown that $\mathbb{R}$ is a complete ordered field satisfying axioms A1 – A11 of Chapter 4 (prove as much as you can up to the point of exhaustion!). We only show here that $\mathbb{R}$ satisfies the completeness axiom.

For cuts $\boldsymbol{A}$ and $\boldsymbol{B}$ we say $\boldsymbol{A} \leq \boldsymbol{B}$ if and only if $\boldsymbol{A} \subset \boldsymbol{B}$. Let $\boldsymbol{S}$ be a nonempty set of real numbers (cuts). Let $\boldsymbol{S}$ be bounded above by say $\boldsymbol{A}$, that is, $\boldsymbol{P} \leq \boldsymbol{A}$ for $\boldsymbol{P}$ in $\boldsymbol{S}$. Consider $\bigcup\{\boldsymbol{P} \mid \boldsymbol{P} \in \boldsymbol{S}\} = \boldsymbol{T}$. Then $\boldsymbol{T}$ is a cut. Clearly, $\boldsymbol{P} \leq \boldsymbol{T}$ for all $\boldsymbol{P}$. Moreover, if $\boldsymbol{P} \leq \boldsymbol{C}$ for all $\boldsymbol{P}$, then $\boldsymbol{P} \subset \boldsymbol{C}$ for all $\boldsymbol{P}$, and so $\boldsymbol{T} \leq \boldsymbol{C}$. So $\boldsymbol{T}$ is a least upper bound for the set of real numbers $\boldsymbol{S}$.

Chapter 6

Zorn's Lemma

Zorn's Lemma is a principle of set theory that plays a crucial role in the proof of many important theorems in advanced mathematics. Among them are the Hahn-Banach Theorem and the Spectral Theorem in analysis, and the Tychonov Product Theorem in topology. In spite of the great importance of Zorn's Lemma, the content of the lemma is quite unintuitive. For this reason the motivation of Zorn's Lemma introduced in this chapter is not as strong as we would like. However, be secure in the knowledge that you will find Zorn's Lemma indespensible in the further study of mathematics.

We shall lead up to the statement of Zorn's Lemma with some preliminary definitions and examples.

6.1 Preliminary Definitions

We begin with the definitions of three types of ordering of a set.

Definition 6.1.1

A partially ordered set is a pair $(P, \ll)$ such that P is a set and $\ll$ is a relation on P such that:

1) for all x in P, $x \ll x$,
2) for all x and y in P, if $x \ll y$ and $y \ll x$ then $x = y$, and
3) for all x, y, and z in P, if $x \ll y$, and $y \ll z$ then $x \ll z$.

We call a partially ordered set a poset for short. We also say that $\ll$ partially orders P. If $x \ll y$ we say x is less than or equal to y, or y is greater than or equal to x. •

We call 2) antisymmetry. Of course, 1) and 3) are reflexivity and transitivity, respectively. Thus, a poset is a set together with a reflexive, antisymmetric, and transitive relation on it.

Example

A common example of a poset is the power set $\mathcal{P}(A)$ of a set A. The power set $\mathcal{P}(A)$ is partially ordered by inclusion ($\subset$) . •

It is important to notice that in a poset we cannot always compare two elements. In the example of the power set $\mathcal{P}(A)$, if S and T are in $\mathcal{P}(A)$ we cannot necessarily assert that $S \subset T$ or $T \subset S$. This leads to our next definition.

Definition 6.1.2

A totally ordered set $(P, \leq)$ is a poset such that if x and y are in P then $x \leq y$ or $y \leq x$. •

Thus, a totally ordered set is a poset in which every pair of elements can be compared.

Example

Consider $(\mathbb{R}, \leq)$ where $\mathbb{R}$ is the set of real numbers and $<$ is the usual ordering of the real numbers. •

Definition 6.1.3

A subset S of a poset P is called a chain if S is totally ordered by the partial order on P. •

Example

Let $A = \{1,2,3,4\}$, and let $P = \mathcal{P}(A)$. Let P be partially ordered by inclusion as before. Let $S = \{\{1\}, \{1, 3\}, \{1, 3, 4\}\}$, then S is a chain. But, $T = \{\{2\}, \{2, 4\}, \{1, 4\}\}$ is not a chain. •

Definition 6.1.4

A totally ordered set $(P, \leq)$ is said to be well-ordered if and only if every nonempty subset has a least element. We say that $x \in S$ is the least element in S if for all $y \in S$, $x \leq y$. •

Examples

a) The natural numbers, **N**, together with the usual ordering is well-ordered.
b) The integers, **Z**, together with the usual ordering is not well-ordered. •

The reader who thinks that **N** is essentially the only example of a well-ordered set is in for a big surprise in Section 6.3

Definition 6.1.5

Let W be a well-ordered set totally ordered by $\leq$, and let $w \in W$. Then the set $W/w = \{s \mid s \in W \text{ and } s < w\}$ is called an initial segment of W. •

Definition 6.1.6

If $(P, \ll)$ is a poset and x is in P we say x is a maximal element of P if and only if there is no y in P such that $x \ll y$ and $x \neq y$. •

Example

Let $P = \{1, 2, 3, 4, 5, 6\}$. Define $\ll$ as follows. Let $1 \ll 2$, $2 \ll 3$, $1 \ll 3$, and $4 \ll 6$. Of course for all x in P set $x \ll x$. Then 3, 5, and 6 are maximal. The rest are not. •

Example

Let P be the set of all open intervals in the real numbers $\mathbb{R}$ that have length strictly less than 1. We define $(a,b) \ll (c,d)$ to mean $(a,b) \subset (c,d)$. This defines a poset. This poset has no maximal elements. •

We remark that a maximal element in a totally ordered set is unique if it exists. Moreover, if y is maximal in a totally ordered set then for all x in the set $x \leq y$. Naturally, we call such a y the maximum of the totally ordered set.

One defines minimal and minimum elements in the obvious way.

Definition 6.1.7

If S is a subset of a poset P, and if $b \in P$, and if for all $x \in S$, $x \ll b$, then we say b is an upper bound of S. •

Exercises for Section 6.1

1. Finite posets can be represented in the following way. On a sheet of paper mark minimal elements as dots along the bottom of the page. For each minimal element find the next elements immediately greater, if any. Mark these above the least element and join with a line. In short, if $x \ll y$, position y on a level above x and join with a line. Continue until the poset is diagrammed. Diagram posets with 1, 2, 4, and 10 elements. Not all chains please!

2. Give an example of a relation that is antisymmetric and transitive but not reflexive.

3. Give an example of a poset that is not totally ordered, has a maximum, but has no minimum.

4. In a poset $\boldsymbol{P}$, is the union of two chains a chain?

5. In a poset $\boldsymbol{P}$, is the intersection of two chains a chain?

6. Let $\boldsymbol{W}$ be a well-ordered set. Let $\boldsymbol{S} \subset \boldsymbol{W}$. Suppose that the least element of $\boldsymbol{W}$ is in $\boldsymbol{S}$, and that for every $\boldsymbol{w} \in \boldsymbol{W}$, if $\boldsymbol{W}/\boldsymbol{w} \subset \boldsymbol{S}$ then $\boldsymbol{w} \in \boldsymbol{S}$. Prove that $\boldsymbol{S} = \boldsymbol{W}$. This is called the transfinite induction principle.

7. Let $\boldsymbol{W}$ be well-ordered and infinite, and suppose that every initial segment is finite. Prove that $\boldsymbol{W}$ is a whole number system in a natural way. Hint, let $S(x)$ be the least element of $\{y \mid y > x\}$.

8. a) Find a well-ordered set that has a maximum element.
   b) Give an example of a well-ordered set in which a nonempty initial segment does not have a maximum element.

6.2 Zorn's Lemma

In this section we shall address the question of when we can be sure that a poset has a maximal element.

In a finite poset $\boldsymbol{P}$ it is easy to find a maximal element. Start at any point and move "upward" (see Exercise 1 in Section 6.1) until a maximal point is reached. In an infinite poset things are not so simple. First, the poset may not have a maximal element. Second, even if it does, one cannot necessarily find a maximal element by starting at a point and moving up, for the chain along which one choses to move may not have a maximum! However, chances would be better if the poset had the property that every chain had an upper bound. One might try the following. Start at some point. If there is nothing greater we are done. If there is, move to that new point and survey the situation again. The hope is to continue this process until the maximum at the end of a chain is reached. Unfortunately, this process is badly flawed. There is no reason why this discrete step by step process should ever produce the desired maximal point. For example, consider the real numbers less than or equal to 1. Start at 0, say. There is a number greater than 0 but less than 1. Move to this new number and repeat the process. A maximal chain is not necessarily produced. Yet, every chain has an upper bound, namely 1.

In spite of the flawed reasoning above it is nevertheless true that a poset in which every chain has an upper bound has a maximal element. We state this as a theorem which we will prove in the next section. In the proof the reader will see that we will use a transfinite version of the flawed procedure just described.

Theorem 6.2.1 (Zorn's Lemma)

If $(P, \ll)$ is a nonempty poset in which every chain has an upper bound then P has a maximal element.

We conclude this section with an application of Zorn's Lemma.

The Existence of a Hamel Basis

We are going to show that there is a subset B of the real numbers $\mathbb{R}$ such that for any r in $\mathbb{R}$, there is a natural number n, rational numbers $q_1, \ldots, q_n$, and elements $x_1, \ldots, x_n$ in B such that $r = q_1x_1 + \ldots + q_nx_n$ and such that any other such representation of r is the same except for the order of the terms.

The following definition gives a condition that guaranties the uniquness just described.

Definition 6.2.2

Let S be a nonempty subset of $\mathbb{R}$. We say S is a linearly independent set over the rational numbers Q if and only if for any $x_1, \ldots, x_n \in S$, and for any $q_1, \ldots, q_n \in \mathrm{Q}$ if $q_1x_1 + \ldots + q_nx_n = 0$ then $q_1 = q_2 = \ldots = q_n = 0$. We call an expression of the form $q_1x_1 + \ldots + q_nx_n$ a linear combination. If every element of $\mathbb{R}$ can be written as a linear combination of elements of S, and if S is linearly independent over Q, we say S is a basis for $\mathbb{R}$ over Q. •

It is easy to see that two linear combinations of the elements of a linearly independent set S are equal if and only if they are the same except for order.

The reader familiar with the notion of a vector space will recognize that we are thinking of $\mathbb{R}$ as a vector space over the rational numbers Q. Moreover, Definition 6.2.2 is the standard definition of a linearly independent subset and basis of a vector space. Therefore, readers familiar with vector spaces can easily modify the proof given below to prove that every vector space has a basis.

Theorem 6.2.3

$\mathbb{R}$ has a basis over Q.

Proof:
Consider the family P of all linearly independent subsets of $\mathbb{R}$. The family is not empty since singletons are linearly independent. Order this family by set inclusion. This makes the family P into a poset. Now consider any chain in P, call it C. We show that $\bigcup C$, the union of all sets in C, is a linearly independent set. Suppose $x_1, \ldots, x_n \in \bigcup C$ and suppose $q_1, \ldots, q_n$ are rationals such that $q_1x_1 + \ldots + q_nx_n = 0$. Since C is a chain there is a linearly independent set S in C to which all the x_i belong. (Why?) Thus, $q_1 = q_2 = \ldots = q_n = 0$. Therefore, $\bigcup C$ is an upper bound for the chain C. So every chain has an upper bound. Therefore, by Zorn's Lemma there is a maximal element B in P. So, B is a maximal linearly independent subest of $\mathbb{R}$. We leave it to the reader to show that B is a basis of $\mathbb{R}$. •

A basis B of the reals over the rationals is called a Hamel basis.

Exercises for Section 6.2

1. Prove that two linear combinations of the elements of a linearly independent set S are equal if and only if they are the same except for order.

2. Complete the proof of Theorem 6.2.3.

3. Consider functions $f: \mathbb{R} \to \mathbb{R}$ of the form $f(x + y) = f(x) + f(y)$ for all x and y in $\mathbb{R}$. If m is a real number then f defined by $f(x) = mx$ is of this form. Are all functions of the form $f(x + y) = f(x) + f(y)$ of the form $f(x) = mx$? This is a challenging problem. Hint: Use a Hamel basis and think about the fact that $f(2x) = 2f(x)$.

6.3 A Proof of Zorn's Lemma

In this section we prove Zorn's Lemma. To do this we shall need The Axiom of Choice, so we review this axiom now (See Chapter 3, Section 3.6).

Let $\mathcal{F}$ be a nonempty family of nonempty sets. It seems plausible that we can form a new set from $\mathcal{F}$ by choosing one element from each set in $\mathcal{F}$. The Axiom of Choice is the official assertion that we can make such new sets.

The fancy way to express such a choice is to say we have a function $h : \mathcal{F} \to \bigcup \mathcal{F}$ such that if S is in $\mathcal{F}$ then $h(S)$ is in S. So $h(S)$ is the element chosen in S. We call the function h a choice function for $\mathcal{F}$. The Axiom of Choice says that for every nonempty family of nonempty sets $\mathcal{F}$ there is a choice function.

Before proving Zorn's Lemma we remark that Zorn's Lemma implies The Axiom of Choice. Thus, The Axiom of Choice and Zorn's Lemma are equivalent. The proof of this is the final exercise of this book.

In order to prove Zorn's Lemma we shall first prove the following famous and surprising theorem.

Theorem 6.3.1 (The Well-Ordering Princple)

Every set can be equipped with a relation that totally orders the set and which is a well-ordering as well. In short, every set can be well-ordered.

Proof:

Let S be a set. Let $h : \mathcal{P}(S) \to \bigcup \mathcal{P}(S)$ be a choice function for $\mathcal{P}(S)$.

Let W be subset of S, and suppose W is well-ordered by some total ordering of W. We say W is h-well-ordered if and only if for all w in W, $h(S - W/w) = w$.

Now let $w_0 = h(S)$, then the set $W_0 = \{w_0\}$ is h-well-ordered. Note that there is only one order relation on this set, namely, $w_0 \leq w_0$. The first element of any h-well-ordered set W is w_0. Indeed, if w is the first element of W then $W/w = \phi$. So, $w = h(S - \phi) = h(S) = w_0$.

The rest of the proof consists in showing the following. First, two h-well-ordered subsets W_1 and W_2 of S are equal and have the same ordering, or one is an initial segment of the other. Second, we shall show that the union of all h-well-ordered subsets is h-well-ordered. Finally, we show that this union is S.

Let W_1 and W_2 be h-well-ordered subsets of S. We use a proof by contradiction to show that they are equal as well-ordered sets, or one is an initial segment of the other. Suppose W_1 and W_2 are not equal as sets or they do not have the same ordering. And also suppose that neither is an initial segment of the other.

Now let

$$T = \{w \mid w \in W_1 \cap W_2, \text{ and } W_1/w = W_2/w \text{ as sets}\}.$$

Suppose $T = W_1$. Suppose s and t are in W_1, then s and t are in W_2. We may assume $s <_1 t$ ($\leq_i$ is the order on W_i). Now $W_1/t = W_2/t$, and so $s \in W_1/t$. But then $s \in W_2/t$, and so $s <_2 t$. Therefore, W_1 is a subset of W_2 ordered by the relation of W_2. Now, $W_1 \neq W_2$, so let x be the least element in $W_2 - W_1$. We claim that $W_1 = W_2/x$. If $y \in W_2$ and $y <_2 x$ then $y \in W_1$ by the way x was chosen. Thus, $W_2/x \subset W_1$. If $y \in W_1$ then $y \in T$. If $x <_2 y$ then $x \in W_2/y$, and so $x \in W_1/y$. But that again contradicts the choice of x. Thus, $W_1 \subset W_2/x$, and so $W_1 = W_2/x$. Therefore, W_1 is an initial segment of W_2. But that contradicts the contradiction premise. So $T \neq W_1$, and similarly $T \neq W_2$.

Now, $W_1 - T$ and $W_2 - T$ are nonempty and well-ordered, and so they have least elements w_1 and w_2. Suppose $t \in W_1/w_1$, then $t <_1 w_1$, and so $t \in T$. Therefore, $t \subset W_2$. We now show that $t <_2 w_2$ in W_2.

Now, $t \neq w_2$, so if $w_2 <_2 t$ then $w_2 \in W_2/t$. But, $W_2/t = W_1/t$. So, $w_2 \in W_1/t$. Thus, $w_2 <_1 w_1$, since $t <_1 w_1$. But then w_1 is not the least element in $W_1 - T$. So, $t <_2 w_2$.

Thus, $W_1/w_1 \subset W_2/w_2$ and similarly $W_2/w_2 \subset W_1/w_1$, and so $W_1/w_1 = W_2/w_2$. But then $w_1 = h(S - W_1/w_1) = h(S - W_2/w_2) = w_2$. Therefore, w_1 (and w_2) is in T by the way T is defined. This is impossible.

We may therefore assume that any h-well-ordered sets W_1 and W_2 are equal and have the same ordering, or one is an initial segment of the other.

Let W be the union of all h-well-ordered subsets of S. Note that W is totally ordered. Indeed, if x and y are in W then x and y are in h-well-ordered sets W_1 and W_2, but they are equal and have the same order, or one is a segment of the other. So x and y lie in a well-ordered set, and so we take the ordering of x and y to be that of an h-well-ordered set to which they belong. Obviously, the order is well-defined.

We show that W is well-ordered. Let A be a nonempty subset of W. Choose any p in A. Then p is in some h-well-ordered set W_1. So $W_1 \cap A$ is not empty. Let x be the least element of $W_1 \cap A$ in W_1. Suppose $y \in A$, and $y < x$. Then x and y belong to an h-well-ordered set W_2. Now $W_1 = W_2$ or one is a segment of the other. In any case $y \in W_1$. But then $y \in W_1 \cap A$, and this contradicts the choice of x. So $x \leq y$, and x is the least element of A.

Observe that if w is in W then w is in some h-well-ordered set W_1. We show that $W/w = W_1/w$. Indeed, if $x \in W_1/w$ then $x < w$, so $x \in W/w$. Conversely, if $x \in W/w$ then $x < w$ and x and w are in some h-well-ordered set W_2. Again W_1 and W_2 are equal or one is a segment of the other. In any case x and w are in W_1. Thus, $x \in W_1/w$. So, $h(S - W/w) = h(S - W_1/w) = w$. Thus, W is h-well-ordered.

Finally, if $W \neq S$, set $y = h(S - W)$. Then note that we can well-order $\underline{W} = W \cup \{y\}$ by setting $x \leq y$ for all x in W. Now $\underline{W}/y = W$, and so $h(S - \underline{W}/y) = h(S - W) = y$. Thus, $\underline{W}$ is h-well-ordered, so $\underline{W} \subset W$. But that is impossible. Hence, $W = S$, and S is well-ordered. •

The Proof of Zorn's Lemma

Let $(P, \ll)$ be a poset in which each chain has an upper bound. Now P is a set, so P has a well-ordering by, say, $\leq$. In order to avoid confusion we will use $\underline{P}$ when we are thinking of P as well-ordered. We use Greek letters where possible in $\underline{P}$. It will also help if we index the elements of P with the elements of $\underline{P}$. Thus, $P = \{p_\alpha \mid \alpha \in \underline{P}\}$, indeed take $p_\alpha = \alpha$ if you like.

Let $p_0 = \alpha_0$ be the least element of $\underline{P}$ in the well-ordering. We define a function $f : \underline{P} \to P$ as follows. Let $f(\alpha_0) = p_0$. Let $\alpha \in \underline{P}$. Assume that $f(\gamma)$ has been defined for all $\gamma < \alpha$. Consider $S_\alpha = \{f(\gamma) \mid \gamma < \alpha\}$. If there is a q such that $f(\gamma) \ll q$ and $f(\gamma) \neq q$ for all $f(\gamma)$, we call q a strict upper bound for S_α. If S_α has a strict upper bound in the partial ordering of P, then by the well-ordering of $\underline{P}$ there is a least index δ such that p_δ is a strict upper bound in P. In this case we set $f(\alpha) = p_\delta$. If S_α has no strict upper bound we set $f(\alpha) = p_0$. This defines f, as we now prove.

Suppose f is not defined for some element of $\underline{P}$, then the set of all such elements at which f is not defined is not empty. So it has a least element, say β. But then f is defined on $\underline{P}/\beta$, and so f is defined at β by the above construction. This is a contradiction, and so f is defined on all of $\underline{P}$. This is transfinite induction at work.

Note that, by the way f is defined, if $f(\alpha) = p_\delta$ then $\delta = \alpha_0$ or $\delta \geq \alpha$.

Let $C = \{f(\alpha) \mid \alpha \in \underline{P}\}$. We show that C is a chain in $(P, \ll)$. Pick $f(\alpha_1)$ and $f(\alpha_2)$ in C. We may assume $\alpha_1 < \alpha_2$ or there is nothing to prove. Now $f(\alpha_2)$

is p_0 or it is not. If $f(\alpha_2) = p_0$ then $\{f(\gamma) \mid \gamma < \alpha_2\}$ has no strict upper bound for p_0 is in all the S_α. Thus, $f(\alpha_1) = p_0 = f(\alpha_2)$. If $f(\alpha_2)$ is not p_0 then $f(\alpha_2)$ is a strict upper bound in $(P,\ll)$ of the set $\{f(\gamma) \mid \gamma < \alpha_2\}$. But $f(\alpha_1)$ is in this set, so $f(\alpha_1) \ll f(\alpha_2)$. Therefore, C is a chain.

By hypothesis C has an upper bound, say b. Suppose b is not maximal, then there is an element p_λ in P such that $b \ll p_\lambda$ and $b \neq p_\lambda$. Consider $S_\lambda = \{f(\gamma) \mid \gamma < \lambda\}$. Then p_λ is a strict upper bound of this set. Now let δ be the least index such that p_δ is a strict upper bound. In this case $f(\lambda) = p_\delta$. But then $\delta \geq \lambda$ since $\delta \neq \alpha_0$. However, $\delta \leq \lambda$, and so $\delta = \lambda$. But then, $f(\lambda) = p_\lambda$, and that is impossible. •

Exercise for Section 6.3

Prove that Zorn's Lemma implies The Axiom of Choice. Thus, The Axiom of Choice, The Well-Ordering-Principle, and Zorn's Lemma are all equivalent. The following outline will help. Let $\mathfrak{F}$ be a nonempty family of nonempty sets. Let $\mathcal{F}$ be the family of a functions f with $\mathrm{Dom}(f) \subset \mathfrak{F}$ and $f(S) \in S$ for all S in $\mathrm{Dom}(f)$. Let $S \in \mathfrak{F}$ and $s \in S$, then $f(S) = s$ defines such function where $\mathrm{Dom}(f) = \{S\}$. Partially order $\mathcal{F}$ by $\subset$. Show that $\mathcal{F}$ satisfies the hypotheses of Zorn's Lemma.

INDEX

Q

R

S

T

U

W

Z